Topological Social Choice

Springer
Berlin
Heidelberg
New York
Barcelona
Budapest
Hong Kong
London
Milan
Paris
Santa Clara
Singapore
Tokyo

Geoffrey M. Heal (Ed.)

Topological Social Choice

With 45 Figures

Springer

Professor Geoffrey M. Heal
Columbia University
Columbia Business School
405 Low Library (Mail Code 433)
535 West 116th. Street
New York, NY 10027, USA

ISBN-13:978-3-642-64599-0 e-ISBN-13:978-3-642-60891-9
DOI: 10.1007/978-3-642-60891-9

First published as Special Issue (Volume 14, Number 2, 1997)
in Social Choice and Welfare

ISBN 3-540-63425-8 Springer-Verlag Berlin Heidelberg New York

Cataloging-in-Publication Data applied for
Die Deutsche Bibliothek - CIP-Einheitsaufnahme
Heal, Geoffrey M.:
Topological social choice / Geoffrey M. Heal. - Berlin ; Heidelberg ;
New York ; Barcelona ; Budapest ; Hong Kong ; London ; Milan ;
Paris ; Santa Clara ; Singapore ; Tokyo : Springer, 1997
ISBN 3-540-63425-8

Hardcover design: Erich Kirchner, Heidelberg
SPIN 10642579 42/2202-5 4 3 2 1 0 – Printed on acid-free paper

Preface

The origins of this volume can be traced back to a conference on "Ethics, Economic and Business" organized by Columbia Business School in March of 1993, and held in the splendid facilities of Columbia's Casa Italiana. Preliminary versions of several of the papers were presented at that meeting. In July 1994 the Fields Institute of Mathematical Sciences sponsored a workshop on "Geometry, Topology and Markets": additional papers and more refined versions of the original papers were presented there. They were published in their present versions in Social Choice and Welfare, volume 14, number 2, 1997.

The common aim of these workshops and this volume is to crystallize research in an area which has emerged rapidly in the last fifteen years, the area of topological approaches to social choice and the theory of games. The area is attracting increasing interest from social choice theorists, game theorists, mathematical economists and mathematicians, yet there is no authoritative collection of papers in the area. Nor is there any survey or book to give a perspective and act as a guide to the issues in and contributions to this new area. One of the two aims of this volume is in some measure to play this role: the other aim is of course to present interesting and surprising new results.

Some of the papers that follow provide an overview of the field. These include my own paper with which the volume starts ("Social choice and resource allocation: a topological perspective), and the paper by Paras Mehta ("Topological methods in social choice: an overview"). Chichilnisky's paper ("Market arbitrage, social choice and the core") functions both as a survey and as a source of remarkable new results, as does that of Baryshnikov ("Topological and discrete social choice: in search of a theory"). The remaining papers contain striking new contributions to the area.

I am very grateful to Graciela Chichilnisky, the originator of this field, for support and assistance in producing this volume, and to Maurice Salles, the energetic and imaginative editor of Social Choice and Welfare, who was a valuable participant in the workshops and who encouraged me in the production of the special issue of Social Choice and Welfare on which this book is based. Springer Verlag, in the person of Dr. Werner Mueller, have been unfailingly helpful and supportive in producing both the journal issue and the book. Without financial support from Columbia Business School, the Fields Institute for Mathematical Sciences and the Mathematical Sciences Division of the US National Science Foundation, this project might never have started.

Geoffrey Heal
New York, April 1997

Contents

Social choice and resource allocation: a topological perspective

Geoffrey Heal

Program on Information and Resources and Graduate School of Business, Columbia University, New York, NY 10027, USA

Received: 25 May 1995/Accepted: 23 April 1996

1. Topological social choice

This volume brings together a collection of papers in an area which has evolved rapidly in recent years. This is the area of continuous or topological approaches to the study of social choice, and the analysis of what have come to be known as "Chichilnisky rules". Initiated by Chichilnisky in a paper in *Advances in Mathematics* in 1980 [14], the field has quickly evolved from a separate and distinct area of social choice to one which is integrated into the mainstreams of both social choice and the general equilibrium approach to markets, and which in turn has come to integrate and bridge these previously disparate areas, revealing deep and unexpected connections between them. Many of the papers in this volume were first presented either at a conference on "*Ethics, Economics and Business*" held at Columbia Business School in March 1993, or at a workshop on "*Geometry, Topology and Markets*" held at the Fields Institute for Mathematical Sciences in July 1994. These two meetings were important in catalyzing the growth of this area. My aim in this introduction is to place the papers in this volume in the context of the development of the area in general, and then to give some personal reflections and observations on the growth of this area, its future, and its role in economic theory. Three of the sections of this introduction, "Topological Social Choice", "Incentives", "Infinite Populations", match the three sections into which the papers that follow have been organized. Section 2 of this introduction, on "Social Choice and Resource Allocation", surveys recent results on the relationship between social choice theory and resource allocation theory, a field covered in more detail in Chichilnisky's paper [18] in Sect. 1. This is such a new and fascinating area that it seems to merit specific comments. The final section contains comments and reflections on the contribution made by this area, and on its future role.

1.1. Chichilnisky's formulation

Chichilnisky's original reformulation of the social choice problem differed from Arrow's [1] in one crucial and strategic respect. Rather than working with a discrete choice space, and treating preferences as orderings over this, i.e., as rankings of a discrete set of objects, she made an assumption conventional in the rest of economic theory and took the choice space to be Euclidean, with preferences given by the indifference surfaces familiar from consumer theory. This immediately placed her version of the social choice problem close to classical resource allocation theory, in that she worked with the same categories of objects. It also enabled her to focus on the problem a set of mathematical tools that are more powerful than those of discrete mathematics, which had previously characterized research methods used in social choice. Within this framework she asked the following question:

> When is there a social choice rule (a map from preference profiles to preferences) which is continuous, anonymous, and which respects unanimity?

Continuity is a conventional axiom in resource allocation theory, and indeed in most applications of mathematics, but is nevertheless a different type of axiom from those used by Arrow and his followers, although Baryshnikov in this volume suggests that Arrow's axiom of independence of irrelevant alternatives plays a similar role and in fact implies continuity once the Arrow problem is embedded in a more structured framework. However, anonymity and respect of unanimity are of the same type as used previously, and are relatively weak. Anonymity is an "equal treatment" axiom: what matters is not who votes, but which way they vote. In particular it implies the widely-used non-dictatorship axiom. Respect of unanimity is a very weak condition and simply means that if everyone in society has the same overall preference ordering, then society adopts this common ordering as its own. This is implied by, but does not imply, the Pareto axiom, according to which if all agents prefer alternative A to alternative B, then society must rank A above B.

Within this framework, let the space of preferences on a Euclidean choice space be $\mathscr{P}$, and the number of agents be k. Then one is asking for a map Φ from $\mathscr{P}^k$ to $\mathscr{P}$ which is continuous, the identity on the diagonal[1] $D = \{p_1, \dots, p_k \colon p_i = p_j \ \forall i,j\}$ and invariant under permutations of its arguments[2], so that $\Phi\ (p_1 p_2, \dots, p_\pi) = \Phi(p_{\pi 1}, p_{\pi 2}, \dots, p_{\pi k})$ where $\pi 1, \dots, \pi k$ is a permutation of the integers 1 to k.

This is an elegant framework admitting analysis by topological methods, which show that in general there is no such rule. This is *Chichilnisky's impossibility theorem*: "*There exists no social choice rule Φ from $\mathscr{P}^k$ to $\mathscr{P}$ which is continuous, anonymous and which respects unanimity*". In fact her argument in [14] is that if there were such a rule Φ, then there would in particular be a map ϕ from $\mathscr{S}^k$ to $\mathscr{S}$ which has the same properties, where $\mathscr{S}$ is the $n-1$-dimensional sphere in the choice space $\mathfrak{R}^n$. Chichilnisky shows that there is no such map on spheres, and hence there is no social choice rule of the

---

[1] This is respect of unanimity.
[2] This is anonymity.

desired type. This framework, and the mathematical arguments and concepts which support it, are surveyed in the paper by Mehta [33] in this volume. This paper by Saari [36] also reviews the structure of social choice theory from a topological and geometric perspective.

This simple and intuitive framework leads to deep results in the area of social choice, and also, more surprisingly, to the development of a very fundamental connection between social choice theory, and the theory of competitive equilibria and cores in exchange economies.

1.2. The choice of topology

A topic on which there has been some confusion, is the role played by the choice of a topology on the space of preferences in Chichilnisky's work. Her basic impossibility theorem is: "*There exists no social choice rule Φ from $\mathscr{P}^k$ to $\mathscr{P}$ which is continuous, anonymous and which respects unanimity*". In her original proof [14], the only condition needed on the topology on the space of preferences for the validity of this result, is that the space of linear preferences on the Euclidean choice space should inherit from this topology the "usual" Euclidean topology. The point here is that linear preferences are a particularly simple type of preference: they can be represented by unit vectors given by their normalized gradient vectors, and so have a natural topology, the Euclidean topology on vectors in finite dimensional spaces. The set of linear preferences can thus be identified with the set of unit vectors in the choice space $\mathfrak{R}^n$, and so with the unit sphere $\mathscr{S}^{n-1}$ of dimension $(n-1)$. Chichilnisky's original argument works with any topology on the overall preference space which reduces to this topology when one considers the subspace of linear preferences. Any reasonable topology, certainly any topology which has been used on preference spaces, satisfies this condition[3].

---

[3] It is satisfied, for example, both by the smooth topology and also by the closed convergence topology. There is a relatively straightforward explanation of the role played by the condition that the space of linear preferences should inherit from the topology on the overall preference space the "usual" Euclidean topology. We seek a continuous map $\Phi: \mathscr{P}^k \to \mathscr{P}$ which is the identity on the diagonal of $\mathscr{P}^k$, and invariant under permutations of its arguments. Consider the space $\mathscr{L}$ of linear preferences on the choice space. Clearly $\mathscr{L}$ is a subspace of $\mathscr{P}$, and the required map Φ would induce a map with the same properties from $\mathscr{L}^k \to \mathscr{P}$. Let $\mathscr{I}$ be the map which includes $\mathscr{L}$ in $\mathscr{P}$. So the composition $\mathscr{I} \circ \Phi$ maps from $\mathscr{L}^k \to \mathscr{P}$. Now consider the map which associates any social preference with its gradient vector at a point, any point, in the choice space. This is the evaluation map: denote it $\mathscr{E}$. This maps social preferences into linear preferences on the choice space. So the overall composition $\mathscr{I} \circ \Phi \circ \mathscr{E}$ sends k-tuples of linear preferences into a linear preference. In this operation, the k-tuple of linear preferences is an element of the overall preference space. If the subspace of linear preferences has the usual Euclidean topology, one has constructed a continuous map from the k-fold product of spheres to the sphere which is the identity on the diagonal of $\mathscr{P}^k$, and invariant under permutations of its arguments. But by analysis of the homotopy groups of these spaces Chichilnisky shows this to be impossible: hence her theorem.

Subsequently it has proved possible to dispense with even this limitation on the topology. Chichilnisky and Heal [21] establish that contractibility[4] of the space of preferences is both necessary and sufficient for the existence of Chichilnisky rules. The Chichilnisky–Heal theorem is as follows: "*Let $\mathscr{P}$ be a para-finite CW complex*[5]*. Then there exists a function Φ from $\mathscr{P}^k$ to $\mathscr{P}$ which is continuous, anonymous and respects unanimity if and only if $\mathscr{P}$ is contractible*". This result implies Chichilnisky's 1980 result, and is true whatever the topology on the preference space, provided that continuity of the social choice rule and continuity in defining contractibility is with respect to the topology on the space of preferences. In this sense it is analogous to statements with which economists are more familiar such as fixed point theorems: "*A continuous function from a convex compact topological space into itself has a fixed point*". This statement is independent of the choice of topology: it is true for any function which is continuous in a topology in which the space is compact, whatever that topology. In fact there is an important mathematical connection between Chichilnisky's impossibility theorem and fixed point theorems: in certain cases her impossibility theorem is equivalent to Brower's fixed point theorem [13, 16]. Related points are made in the paper by Zhou [39] in this volume.

2. Social choice and general equilibrium

One of the most striking results to emerge from topological social choice is the equivalence between Arrow's paradox, Chichilnisky's paradox, the existence of a competitive equilibrium in a general equilibrium exchange economy, and the non-emptiness of the core. This is a remarkable set of relationships:

Arrow Paradox $\Leftrightarrow$ Chichilnisky Paradox $\Leftrightarrow$

Competitive equilibrium $\Leftrightarrow$ Core

This shows that the problem of social choice is an integral part of *the* problem in economics: that rather than being a non-economic problem, as asserted by Samuelson [38] in his review of Arrow's book, all forms of resource allocation (via cooperative games, competitive exchange economies and social choices) are equivalent, at least in terms of the conditions under which they yield interesting outcomes. Social choice has come in from the cold, and can contribute to the resolution of the most classical questions in economics.

This result depends on three pairwise equivalences:

1. the equivalence of the existence of a competitive equilibrium in an exchange economy and the existence of social choice rules in Chichilnisky's sense,

---

[4] A topological space Y is contractible if $\exists y_0 \in Y$, and a continuous function $f: Y \times [0, 1] \to Y$, with $f(y, 0) = y \; \forall y \in Y$ and $f(y, 1) = y_0 \; \forall y \in Y$: see Heal [24] and also the paper by Mehta in this volume.

[5] This is a CW complex with a finite number of cells in each dimension (though not necessarily a finite number of cells in total). See also the paper by Baryshnikov in this volume.

2. the equivalence of the existence of a competitive equilibrium and the non-emptiness of the core, and
3. the equivalence of Chichilnisky's formulation of the social choice paradox[6] and Arrow's original formulation[7].

The first two equivalences are reviewed here in the paper by Chichilnisky [18], and the latter in the paper by Baryshnikov [4] and also for certain cases in the paper by Chichilnisky.

The equivalence of existence of a Chichilnisky rule to the existence of a competitive equilibrium was shown by Chichilnisky in 1991 [12], where she established that a condition called "limited arbitrage" was necessary and sufficient both for the existence of a competitive equilibrium and for the existence of a social choice rule in her sense. Limited arbitrage is a condition requiring similarity of agents' preferences and endowments. Loosely speaking, agents' preferences and endowments satisfy limited arbitrage when the directions in which their utilities increase without bound are similar. Precisely, the condition is the that duals of these sets of directions have a non-empty intersection[8].

In her paper [16], Chichilnisky shows that for very general families of sets, their intersection is non-empty if and only if their union is contractible. In so doing she uses the concept of the nerve of a family of sets, also used by Baryshnikov in this volume. This result enables her to establish that the similarity of preferences which is necessary and sufficient for the existence of a competitive equilibrium, and which is stated in terms of non-empty intersection of a family of sets associated with agents' preferences, is the same as the contractibility condition on the space of preferences shown by Chichilnisky and Heal [21] to be both necessary and sufficient for the existence of Chichilnisky rules. The Chichilnisky–Heal result provided for the first time a clear and complete characterization of the conditions for a resolution of the social choice paradox.

In 1993 Chichilnisky [10] showed that the same conditions are necessary and sufficient for the non-emptiness of the core of an exchange economy: this is an almost immediate corollary of her earlier results.

There are two separate results on the relationship between Arrow's and Chichilnisky's axiomatizations. One is due to Chichilnisky [9, 18]: she shows that for choices over sets of alternatives involving large utility values, the limited arbitrage condition is equivalent to the non-existence of Condorcet triples. From this she can show that limited arbitrage is also equivalent to the non-existence of Arrow's paradox over choices between alternatives involving large utility values. The concept of "choices involving large utility values" is a novel one: Chichilnisky uses this to refer to a property (such as the presence or absence of Condorcet triples) which holds on sets of alternatives whose utility values exceed some lower bound (which of course is less than the supremum of utility values). The focus on large utility values arises because

---

[6] This was in terms of rules which are continuous, anonymous and respect unanimity.
[7] This was in terms of rules which work on a universal domain, and satisfy the Pareto principle, independence of irrelevant alternatives and non-dictatorship.
[8] Limited arbitrage bounds, but not by zero, the gains in utility available from zero cost trades.

limited arbitrage, which was shown by Chichilnisky to be necessary and sufficient for the existence of competitive equilibrium and for the non-emptiness of the core, deals with the asymptotic behavior of preferences.

Baryshnikov completed the picture in 1993 [3] by showing Arrow's formulation of social choice theory to have the same underlying structure as Chichilnisky's: he showed that Arrow's axioms imply the existence of a map from (at the topological level) the product of spheres to the sphere. Recall that Chichilnisky's original argument proceeded by the analysis of a map from the product of spheres to a sphere, and that Chichilnisky's impossibility theorem implies that there is no such map: hence Chichilnisky's result implies Arrow's. Arrow's assumption of independence of irrelevant alternatives ensures that this is a simplicial map and so continuous. The Pareto condition used by Arrow implies respect of unanimity. Baryshnikov's conclusion is that the existence of an Arrow map is fundamentally equivalent to the existence of a Chichilnisky map: both are tied to the existence of symmetric maps on spheres.

So, in summary, Chichilnisky has shown (via a very simple proof) that her formulation and Arrow's are equivalent for choices involving large utility values. Baryshnikov has shown that Arrow's formulation can be reduced to the study of maps from the product of spheres to a sphere, and has the same structure as Chichilnisky's.

The full implications of this set of results are clearly not yet understood. One aspect in particular may lead to interesting new developments: this is the introduction of a formal measure of "social diversity" in Chichilnisky's proofs of equivalence [9, 11]. She shows that competitive equilibria, the core and social choice rules all exist if and only if a certain index of social diversity takes the value one, its minimal value. This index assuming the value unity is equivalent to the limited arbitrage condition. Non-existence of any of these equilibrium concepts can therefore be attributed to the degree of social diversity. This is quite in keeping with the early literature on social choice theory which, from the time of Black [6] onwards, has introduced limitations on the diversity of preferences[9] as a way of ensuring the existence of social choice rules. What is elegant, unexpected and far-reaching in its implications, is that the existence of competitive equilibria and the core also depend on this same concept of diversity.

An important feature of topological social choice is the natural and productive role that it has provided for interesting[10] mathematics in economics: Chichilnisky's original formulation of the problem, Baryshnikov's proof of equivalence between Chichilnisky's and Arrow's formulations, and Chichilnisky's proofs of the equivalences listed above [16], all require the use of techniques from algebraic topology not previously used in economics, but which nevertheless seem well-suited to our subject. The techniques of topology are designed for analysis of problems which are qualitative rather than quantitative in nature, either because numerical representation is not possible or because it is possible only subject to considerable error. They therefore fit

[9] Such as single peakedness (Black) and value restrictedness (Pattanaik and Sen [34]).

[10] Some readers may be inclined to remind me of the Chinese curse "May you live in interesting times!".

well the areas of welfare economics and social choice, and it is encouraging that they have yielded a rich harvest.

3. Incentives

3.1. Manipulability and straightforwardness

One of the most productive developments of social choice theory has been in the area of incentive compatibility and the manipulability of social choice procedures. Gibbard [28] and Satterthwaite [37] observed that the non-existence of social choice rules implied that most social choice rules could be manipulated, in the sense that agents' optimal strategies typically involved revealing preferences other than their true preferences. This opened a productive literature on incentive compatibility, addressing this both in the strict confines of the traditional social choice framework, and in the more economic confines of the selection of levels of provision for public goods, the culmination of which was the Groves–Ledyard mechanism [27] for determining efficient allocations of resources to the provision of public goods and overcoming the classical free rider problem. A substantial part of this literature has been concerned to characterize social choice rules which are "straightforward", in the sense that for all agents facing these rules, revealing their true preferences is a dominant strategy so that they have no incentive to behave strategically or dishonestly. The paper by Chichilnisky and Heal in this volume [22], often cited but not previously published, gives a complete characterization of straightforward rules for a reasonably broad class of preferences, those which are single peaked. It shows that such rules have a remarkably simple structure: they are locally either dictatorial or constant. Clearly rules which are everywhere constant, or everywhere dictatorial, are straightforward: no-one has any incentive to misrepresent. What is surprising, is that all straightforward rules are obtained by "patching together" such functions. The paper by Rasmussen [35] links the literature on implementation to that on topological social choice.

3.2. Strategic dictators

In the topological framework there is a result that is equivalent to that of Gibbard and Satterthwaite: this is that any social choice rules which are continuous and respect unanimity, must admit "strategic dictators" [8]. A strategic dictator is an agent who has the following ability. Whatever the preferences announced by other agents, he or she can always announce a preference such that the outcome of the social choice rule will be his or her true preference. In other words, such an agent always has available a response to any set of moves by the other agents which will produce as the outcome his or her optimal outcome.

A strategic dictator is defined formally as follows. Let $\Psi : \mathscr{P}^k \to \mathscr{P}$ be a continuous social choice rule. Let $s \in \mathscr{P}^k$ be a k-vector of preferences announced by all k agents, and let s_{-i} be the $k-1$ tuple of preferences left when agent i's is removed. Let agent i's true preference be s_i^*. Then agent i is a strategic dictator if for any vector s_{-i} of preferences announced by other agents, there exists a response by agent i, denoted $s_i\ (s_{-i})$, such that the outcome is precisely the true preference of i, i.e. $\Psi(s_i(s_{-i}), s_{-i}) = s_i^*$. Typically, of course, $s_i(s_{-i}) \neq s_i^*$, so that agent i obtains the desired outcome by misrepresentation. So in the Chichilnisky framework, manipulability is again implied by the basis impossibility result. (The papers by Baryshnikov [4] and Mehta [33] in this volume also develop this concept.)

The analysis of strategic dictators requires sophisticated mathematical tools: in mathematical terms it has required the classification of maps from the product of spheres to the sphere, a classical mathematical problem. There are two approaches, via homotopy theory and via homology theory. The former involves showing that any continuous social choice rule which satisfies the Pareto condition and an additional condition of positive association[11] must be homotopic to a map which projects the k-tuple $(p_1, p_2, \ldots, p_k)$ onto one of its components, say the jth. Any such projection map is of course dictatorial, with the jth. agent as the dictator. Two maps $f, g: X \to Y$ between the topological spaces X and Y are homotopic if there exists a continuous map $h: X \times [0,1] \to Y$ such that $h(x,0) = f(x)$, $h(x,1) = g(x)\ \forall x \in X$. In this case the map f can be continuously deformed in to the map g. Chichilnisky [17] formulated the problem of the dictatorial properties of continuous social choice rules in terms of the properties of maps from sphere products to the sphere, and then showed that any continuous map satisfying the Pareto condition and positive association is homotopic to a dictatorial map. In the case of only two agents, the weak association condition is not needed. However, Baryshnikov has shown [5] that in general one cannot dispense with positive association.

An alternative approach is to show that any continuous rule which satisfies only respect of unanimity is homologically equivalent to a projection map and hence to a dictatorial map. Homological equivalence is weaker than homotopic equivalence: if two functions f and g are homologically equivalent, this does not imply that there is a homotopy between them and that one can be deformed into the other. However, this property of being homologically equivalent to a dictatorial rule is still sufficient to establish the existence of a strategic dictator. This is the approach taken in Chichilnisky [8] and in Chichilnisky and Heal [20]: this latter paper considers the case of two agents and in this case the homology argument is equivalent to an argument about the degree of a map from a sphere to a sphere. Chichilnisky shows that any continuous social choice rule which respects unanimity is homologically equivalent to a dictatorial rule and hence that such a rule must be associated with a strategic dictator. If the rule also satisfies the Pareto condition, then the strategic dictator is unique. Koshevoy, in this volume [29], extends Chichilnisky's work on the homotopy equivalence case by dropping the condition of positive association, using only continuity and the Pareto condition, but restricting the choice space to be two dimensional and preferences to be linear. He also extends Chichilnisky's general homological results to the case of infinite populations.

---

[11] Weak positive association (wpac) in the paper by Koshevoy in this volume.

Chichilnisky and Heal [20] begin to explore the implications of these results for game theory. The key point is as follows: if agent i is a strategic dictator, then at a Nash equilibrium of agents' choices of preferences, the outcome will be agent i's true preference. But if there are two distinct agents who are strategic dictators, then the outcome must equal the true preferences of both. This is a contradiction, implying that in this case there will be no Nash equilibrium to the preference revelation game. This game will have an equilibrium only if there is a unique strategic manipulator, in which case the equilibrium will of course be his or her preferred outcome.

4. Infinite populations

It was recognized by Fishburn [26] and Kirman and Sonderman [25] that moving to an infinite population of voters made a fundamental difference to the social choice problem in Arrow's framework. To be precise, they showed that Arrow's paradox can be resolved in this context. Chichilnisky and Heal, in a paper initially circulated in 1979 but published for the first time in this volume [23], studied the structure of continuous social choice rules with infinite populations. They showed that, as in the discrete case, with infinite populations the resolution is in some sense more apparent than real: both Arrow and Chichilnisky show that "desirable" social choice rules are dictatorial, directly or strategically. This is still true with infinite populations of voters, but the dictators are now "invisible": Chichilnisky and Heal show that they are, for example, the limits of sequences of voters, or generalizations of limits via ultrafilters. In fact they provide a characterization of all continuous rules admissable with infinite populations. They also show that although no individual agent is a dictator, if the set of agents (the integers) is given a finite measure, then there exists a coalition of agents of arbitrarily small measure which determines fully the outcome of the social choice rule.

This approach has recently been applied to intertemporal choice, where there is naturally an infinite sequence of generations. Chichilnisky [7] has used an extension of the topological approach to social choice with infinite populations to provide an axiomatization of alternatives to the utilitarian framework to intertemporal welfare economics, and has related this to the concept of sustainable growth. She uses three axioms, a continuity condition, and then two non-dictatorship conditions, non-dictatorship of "the present" and non-dictatorship of "the future". She shows that these axioms imply that infinite utility streams must be evaluated according to two terms, one which is the integral of the stream against a countably additive measure (such as an integral of discounted utilities) and one of which is a purely finitely additive measure, such as the long run average utility level or the lim sup or lim inf of utility levels. This axiomatization, derived from topological social choice theory, therefore justifies placing more weight on the very long run than the standard discounted utilitarian approach, and also justifies the use of a mix of the intertemporal optimization criteria used in optimal growth theory (discounted integral of utilities) and in the theory of repeated games (long run average utility value).

Lauwers [30, 31, 32] has also analyzed intertemporal welfare economics within this framework, focussing on issues arising from alternative interpretations of the anonymity axiom with infinite populations. In the work of Lauwers, and that of Chichilnisky, a natural and important role emerges for purely finitely additive measures defined on consumption or utility sequences. These purely finitely additive measures correspond to the limits and generalized limits which emerge as "invisible dictators" in the analysis of social choice with infinite populations. Lauwers' work is contained in his papers in this volume.

The paper in this volume by Candeal et al. [19] is also in the tradition of social choice rules with infinite populations, but considers the technically more demanding case of a continuum of agents. In this case the interpretation of axioms like anonymity is particularly delicate, and it is difficult both to formalize this and to establish positive results about the existence of rules satisfying it. Candeal, Chichilnisky and Indurain propose a formalization and show that with this approach one can establish a generalization of the Chichilnisky–Heal possibility theorem to the effect that contractibility of the space of preferences is necessary and sufficient for the existence of continuous rules which satisfy anonymity and respect of unanimity.

5. Where have we come from? Where are we going?

As economists, we are concerned to understand the mechanisms by which resources may be allocated, and to provide a basis for evaluating these. In an increasing number of countries, the market is a predominant mechanism. But even in the most market-oriented economy, many dimensions of resource allocation are within the domain of political decision-making, and are therefore determined by social choices. One thinks immediately of defense, of education, of health care, and of infrastructure investments. These account for probably one third of gross national product in most industrial economies. And even in those areas which are the domain of the market, many collective decisions are made concerning the framework within which the market will operate. Examples are telecommunications and broadcasting. In fact, even in the securities business, one of the most competitive of industries and the archetype of capitalist success, many collective decisions determine the laws which regulate the organization and conduct of the industry. So social choices and social decision-making are integral to the agenda of economics: they are important determinants of the pattern of resource allocation in a society. As economists, we therefore need to understand them, and to understand how they relate to other resource-allocation mechanisms. This is why social choice theory matters.

Prior to Chichilnisky's 1980 article, the paradigm for social choice was still that set by Arrow over fourty years ago. This framework had been productive in exploring the robustness of Arrow's original result with respect to alterations of the axiom set, and had shown the basic paradox to be strong. It had also opened up the area of incentive-compatibility. But it seems fair to say that it had been weak in establishing clear general results and a unified perspective on the subject, and in providing links to the rest of economics. Together with the importance of the underlying subject matter, this provided an opening for an alternative approach.

In the early days of the topological approach to social choice, a comment often heard (though to the best of my knowledge never written down) was that this was not *really* social choice in the traditional sense: it was a different and largely mathematical exercise. We now know that to be wrong: the relationships established between Chichilnisky's and Arrow's formulations show that at a deep level they are capturing the same phenomenon, a fundamental aspect of the process of combining individual preferences into a social preferences through democratic institutions. Furthermore, we also know now that the possibilities and limitations facing us for combining individual preferences through democratic institutions are not very different from those available to us when we look at alternative institutions, such as competitive markets or cooperative games. This naturally leads one to ask whether there are other institutions which are more robust in their domain of applicability, and could provide satisfactory resolutions of social choices (including the allocation of resources) in the face of social diversity greater than that which can be managed by the institutions reviewed in this volume. This seems to be a high priority area for research, although a difficult one. One might say that the results to date have led us to understand better what we know, and to see how various parts of this fit together. Now we need to apply this methodology to problems to which we do not yet have answers, perhaps to problems which we have not yet posed.

There is an important difference in emphasis between the two approaches: the combinatorial approach was always taken to imply that it is very hard to find workable and ethically desirable social choice procedures. Arrow's book, for example, is remembered for the impossibility theorem. The topological approach has made social choice seem more attainable: it has shown that for the existence of social choice rules one needs only that the space of agents' preferences be contractible, a relatively mild limitation on the diversity of individual preferences. This approach has also shown us that it is as easy, or as difficult, to find acceptable social choice rules, as it is to find competitive equilibria in exchange economies, or cores of exchange economies. We are accustomed to thinking of this as easy, although in fact early on in the development of general equilibrium theory Arrow provided a rather robust example of an exchange economy with no competitive equilibrium (repeated in Arrow and Hahn [2] and also in Chichilnisky [11]). Of course, there is really no conflict between these interpretations: it is a matter of what the various authors have chosen to emphasize, and also a product of the fact that the combinatorial approach never produced a complete characterization of the conditions needed for the existence of an acceptable social choice rule, so that we never had a clear picture of how weak these could be, whereas such a characterization was provided relatively early on in the development of Chichilnisky's approach.

Perhaps a fair way of characterizing a part of the difference between the traditional and the topological approaches, is in terms of a trade-off between the investment required to enter, and the long-run payoffs available. The traditional approach, relying on combinatorics and logic, needed relatively little investment in specialized techniques on the part of researchers wishing to work in the area: anyone with a certain degree of analytical sophistication could master the techniques and enter the area. But the ultimate payoffs were limited: the full structure of the social choice problem and its relationship to the rest of resource-allocation theory could not be appreciated by using this methodology. In contrast, algebraic topology has set-up costs: its requires an investment of time before it can be used. Once it is understood, however, it gives greater insights into the structure of the area at relatively low incremental cost. It also gives a geometric insight into the structure of social choice theory, as emphasized in the papers of Baryshnikov and Saari that follow. I was struck by the statement in Baryshnikov's conclusion that he started reproving the Arrow theorem topologically because he had no intuitive feeling for the result from a combinatorial argument, and wanted a proof that matched this striking and beautiful result in aesthetic appeal! This matches well my own motives for becoming involved in this area.

There is a greater fixed cost, but a lower marginal cost, associated with the topological approach: a researcher in the field for the long term should therefore make the investment. In some respects, this is a metaphor for what has happened in much of economic theory: economists have tended to stay within the areas which can be understood by traditional mathematical techniques requiring little investment, and have reached or passed the point of diminishing returns in applying these techniques. An interesting analogy is with the area of dynamical systems: difference equations are relatively easy to formulate and to analyze. They have low set-up costs. But real progress in dynamical systems depended upon the introduction of the continuous approach and the use of a technology based on continuous and differentiable systems.

Reviewing the results in this volume, and their predecessors, one cannot avoid the conclusion that the topological approach has yielded very substantial insights into social choice theory and its relationship to resource allocation theory and mainstream economics, and has done so in a relatively short time. Chichilnisky's first paper appeared in 1980, and given its novelty and the technical demands it placed on the reader, it is natural that it was several years before this was understood and used by others. Since then, however, this approach has generated a large literature and has to its credit an elucidation of the conditions on preference diversity under which social choice is or is not possible, a characterization of the social choice rules which are then possible, showing these to be generalized means, a clarification of the relationship with prior work, and an integration of social choice into the mainstream of economics, leading to surprising and fundamental insights into the relationships between different resource allocation mechanisms. It also has insights, whose potential has not yet been exploited, into the theory of non-cooperative games and into the general area of social diversity, its measurement and its role in the functioning of society. Such achievements and potential constitute a validation of the framework which has produced them. Realizing the potential of this approach in the areas of games and of understanding social diversity seems to offer very high rewards.

References

1. Arrow KJ (1951) Social Choice and Individual Values. Wiley, New York
2. Arrow KJ, Hahn FH (1971) General Competitive Analysis. Holden Day, San Francisco
3. Baryshnikov Y (1993) Unifying impossibility theorems: a topological approach. Adv Applied Math 14: 404–415

4. Baryshnikov Y (1997) Topological and discrete social choice: in search of a theory. Soc Choice Welfare 14: 199–209
5. Baryshnikov Y (1994) Strategic dictators need no be homotopic dictators. Paper presented at the conference on Geometry, Topology and Markets at the Fields Institute for Mathematical Sciences
6. Black D (1948) The rationale of group decision-making. J Polit Econ 56: 23–24
7. Chichilnisky G (1994) What is Sustainable Development? Stanford Institute for Theoretical Economics Technical Report No. 65. Published as: An axiomatic approach to sustainable development. Soc Choice and Welfare (1996) 13: 231–257
8. Chichilnisky G (1993) On strategic control. Q J Econ 108: 285–290
9. Chichilnisky G (1994) Social diversity, arbitrage and gains from trade: a unified perspective on resource allocation. Am Econ Rev 84(2): 427–434
10. Chichilnisky G (1994) Limited arbitrage is necessary and sufficient for the existence of competitive equilibrium, the core, and limits voting cycle. Econ Letters
11. Chichilnisky G (1995) Limited arbitrage is necessary and sufficient for the existence of a competitive equilibrium. Econ Theory 5(1): 79–108
12. Chichilnisky G (1991) Markets, arbitrage and social choice. Department of Economics, Columbia University. CORE Discussion Paper No. 9342, 1993
13. Chichilnisky G (1979) Fixed point theorems and social choice paradoxes. Econ Letters 3: 347–351
14. Chichilnisky G (1980) Social choice and the topology of spaces of preferences. Adv Math 37(2): 165–176
5. Chichilnisky G (1982) Social aggregation rules and continuity. Q J Econ 97(2): 337–352
16. Chichilnisky G (1993) Interesting families of sets and the topology of cones in economics. Bull Am Math Society 29(2): 189–207
17. Chichilnisky G (1982) The topological equivalence of the pareto condition and the existence of a dictator. J Math Econ 9: 223–233
18. Chichilnisky G (1997) Market arbitrage, social choice and the core. Soc Choice Welfare 14: 161–198
19. Chichilnisky G, Indurain E, Uriarte JR (1997) Soc Choice Welfare 14:
20. Chichilnisky G, Heal GM (1984) Patterns of power: bargaining and games in two-person situations. J Public Econ 23: 333–349
21. Chichilnisky G, Heal GM (1983) A Necessary and sufficient condition for the resolution of the social choice paradox. J Econ Theory 31: 68–87
22. Chichilnisky G, Heal GM (1997) The geometry of implementation: a necessary and sufficient condition for straightforwardness. Soc Choice Welfare 14: 259–294
23. Chichilnisky G, Heal GM (1997) Social choice with infinite populations: construction of a rule and impossibility results. Soc Choice Welfare 14: 303–318
24. Heal GM (1983) Contractibility and public decision-making. In Pattanaik P, Salles M (eds) Soc Choice Welfare. North Holland, Amsterdam
25. Kirman A, Sonderman D (1972) Arrow's Theorem, Many Agents and Invisible Dictators. J Econ Theory 5: 267–277
26. Fishburn PC (1970) Arrow's impossibility theorem: concise proof and infinite voters. J Econ Theory 103–106
27. Groves T, Ledyard J (1977) Optimal allocation of public goods: a solution to the free rider' problem. Econometrica 45: 783–811
28. Gibbard A (1973) Manipulation of voting schemes: a general result. Econometrica 61: 587–602
29. Koshevoy G (1997) Homotopy properties of Pareto aggregation rules. Soc Choice Welfare 14: 295–302
30. Lauwers L (1997) A note on weak ∞-Chichilnisky rules. Soc Choice Welfare 14: 357–358

31. Lauwers L (1997) Continuity and equity with infinite horizons. Soc Choice Welfare 14: 345–356
32. Lauwers L (1997) Topological aggregation, the case of an infinite population. Soc Choice Welfare 14: 319–332
33. Mehta P (1977) Topological methods in social choice: an overview. Soc Choice Welfare 14: 233–243
34. Pattanaik PK, Sen AK (1969) Necessary and sufficient conditions for rational choice under majority decision. J Econ Theory 1: 178–202
35. Rasmussen H (1997) Strategy-proofness of continuous aggregation maps. Soc Choice Welfare 14: 249–257
36. Saari D (1997) Informational geometry of social choice. Soc Choice Welfare 14: 211–232
37. Satterthwaite M (1975) Strategy-proofness and Arrow's conditions: existence and correspondence theorems for voting procedures and social welfare functions. J Econ Theory 10· 187–217
38. Samuelson PA (1967) Arrow's mathematical politics. In Hook S (ed) Human Values and Economic Policy. N.Y.U. Press, New York
39. Zhou Y (1997) A note on continuous social choice. Soc Choice Welfare 14: 245–248

Market arbitrage, social choice and the core

Graciela Chichilnisky*

Program on Information and Resources, Columbia University,
405 Low Memorial Library, New York, NY 10027, USA

Received: 30 December 1994/Accepted: 12 August 1996

Abstract. This paper establishes a clear connection between equilibrium theory, game theory and social choice theory by showing that, for a well defined social choice problem, a condition which is necessary and sufficient to solve this problem – limited arbitrage – is the same as the condition which is necessary and sufficient to establish the existence of an equilibrium and the core. The connection is strengthened by establishing that a market allocation, which is in the core, can always be realized as a social allocation, i.e. an allocation which is optimal according to an ordering chosen by a social choice rule. Limited arbitrage characterizes those economies without Condorcet triples, and those for which Arrow's paradox can be resolved on choices of large utility values.

1. Introduction

Markets provide a widely used solution to the problem of allocating resources among the members of the economy. A market equilibrium is individually optimal and clears the markets. The efficiency of competitive market allocations is what makes them desirable.

---

* UNESCO Chair in Mathematics and Economics and Director, Program on Information and Resources, Columbia University. email: gc9@columbia.edu. Support from Stanford Institute for Theoretical Economics, NSF Grant SES92-16028, the Sloan Foundation, the Salinbemi Chair at the University of Siena and the Leif Johansen Award at the University of Oslo contributed to the research reported here. The main results of this paper were presented at "Columbia Celebrates Arrow's Contributions" at Columbia University, October 1991, and at invited addresses to the International Congress of Mathematicians Paris, July 4, 1994, a Colloquium at the University of California at Berkeley, March 1993, at the American Economic Association Annual Meetings in Boston, January 3–5 1994, and as a keynote address to the World Congress on Non Linear Analysis in Athens, July 16, 1996.

A different solution to the resource allocation problem is provided by *social choice rules*. These rules derive a social preference as a function of individual preferences in a fair fashion, for example through voting. An allocation which is optimal according to a social preference, is called a *social allocation*.

A third solution to the problem of resource allocations is the *core*: these are allocations from which no coalition would want to deviate.

This paper establishes a clear connection between equilibrium theory, social choice theory and game theory by showing that, for a well defined social choice problem, a condition which is necessary and sufficient to solve the social choice problem – *limited arbitrage* – is the same as the condition which is necessary and sufficient to establish the existence of an equilibrium and the core.[1] The connection is strengthened by establishing that a market allocation, which is in the core, can always be realized as a social allocation.

How does limited arbitrage work? Limited arbitrage is a condition on traders' endowments and preferences. Introduced in [18] it bounds potential gains from trade, and the scope for mutually advantageous reallocation in the economy. In mathematical terms, limited arbitrage is the non-empty intersection of a family of cones. Somewhat surprisingly, this nonempty intersection is the same as a topological condition: the *contractibility* of spaces of preferences. This connection arises from a new result: *a family of convex sets intersects if and only if every subfamily has a contractible union*, proved in Chichilnisky [15, 19]. It has been shown that contractibility is a restriction on the diversity of preferences [44]. Through its connection with contractibility, therefore, limited arbitrage implies a restriction on the traders diversity [22], [23, 26, 28]. Indeed, the central role of limited arbitrage in resource allocation is explained by its connection with social diversity [22], [23], [26], [28]. Social diversity is crucial for resource allocation.

For example, social diversity in the form of contractibility of preferences is necessary and sufficient for the existence of continuous anonymous social choice rules respecting unanimity, Chichilnisky and Heal [33, 18, 12]. Since limited arbitrage is a form of contractibility, this explains the connection between limited arbitrage and social choice. I show below that limited arbitrage is also closely connected with Arrow's impossibility theorem: it is necessary and sufficient to eliminate Condorcet triples – or cyclical behavior – on choices over feasible allocations of utility value which approaches the supremum of utilities, and for solving Arrow's theorem on such choices.[2]

Diversity has a fundamental role in other forms of resource allocation. Indeed, markets exist because people are different. Only when traders have different preferences and endowments do they have a reason to trade. Furthermore a condition which limits diversity, namely limited arbitrage, is necessary

---

[1] These results have been announced and presented in a number of papers starting in 1991: Chichilnisky [18, 16, 17, 22, 21, 23, 24, 20, 26, 31, 27] Chichilnisky and Heal [32]; these papers and the literature are discussed in a comment by Monteiro Page and Wooders [50] and its response Chichilnisky [25].

[2] This is an ordinal result, which is independent from the utility representation of preferences. See also [20, 21, 22, 23].

and sufficient for the existence of a competitive equilibrium in general economies [16, 32, 23, 24, 26]. It is also necessary and sufficient for the uniqueness of equilibrium in strictly regular economies [27]. Elsewhere I showed also that limited arbitrage is necessary and sufficient for the existence of Pareto efficient allocations [28, 29, 30, 16, 24]. Thus a failure of limited arbitrage prevents optimal coalition formation: it leads to an empty core [21, 28]. Diversity is therefore equally important for game theoretical allocations.

These results highlight novel features of resource allocation. For example, market allocations are generally considered to be practical, more so than social allocations. One reason for this is that markets are viewed as having equilibria generally, while social choice theory has always stressed paradoxes and non-existence results. Our results, however, show that this is not accurate. The same restriction on diversity is necessary for the existence of market equilibrium and for the existence of appropriate social choice rules.[3]

A careful examination of the literature illustrates the importance of diversity in resource allocation. The social choice literature has focused on interpersonal diversity as a reason for the non-existence of social choice rules. Several ways of resolving social choice problems have been proposed by limiting the diversity of the individual preferences, a 'domain restriction' on preferences, Black, [8, 9], Pattanaik and Sen [55], and Chichilnisky and Heal [33]. These works propose ways of limiting the diversity of individuals. The issue of existence of universal social choice rules – a problem which in its more general form has no solution – is turned into the question: for what societies can the social choice problem be resolved? Or: how much diversity can a society function with?[4]

The issue of diversity has also been studied in the context of market equilibrium. There are widely used conditions which ensure the existence of a competitive equilibrium. These are also restrictions in social diversity: they require that the endowments of any household are desired, indirectly or directly, by others, so that their incomes cannot fall to zero: examples are McKenzie's *irreducibility* condition [48, 49, 50], and the *resource relatedness* condition in Arrow and Hahn [3]. These conditions are different from limited arbitrage, but nevertheless restrict the diversity of individuals' preferences and endowments. Limited arbitrage provides another limitation in social diversity: it does so parsimoniously, because it is a minimal condition, in the sense of

---

[3] Competitive equilibria are not always available: when the traders endowments and preferences are very different the economy can fail to have a competitive equilibrium. Arrow and Hahn ([3], Chapter 4, 1) provided examples of standard market economies with no competitive equilibrium. Their example highlights the role of interpersonal differences between the traders, differences in their endowments and in their preferences, in preventing the existence of a competitive equilibrium. The failure of existence arises when some traders have zero income, while others wish and can afford unboundedly large allocations which a bounded economy cannot accommodate ([3], Chapter 4, p. 80).

[4] Black's singlepeakedness condition restricts diversity and solves the problem proposed by Condorcet's [34] paradox of majority voting.

being necessary as well as sufficient for existence. In addition, it applies simultaneously to resource allocation by market equilibrium, by social choice[5] and the core. Indeed limited arbitrage measures precisely the amount of diversity acceptable for the functioning of markets, cooperative games and social choice.

There is a different perspective on diversity, one which views diversity as a positive factor in a group's adaptation to its environment. This view contrasts with the role of diversity on resource allocation, in which the functioning of economic institutions imposes limits on diversity. The contrast suggests an open, and perhaps unsettling, question: Are the forms of diversity implied in our economic institutions consistent with humans' successful adaptation to change? Are our economic institutions sustainable?

2. Definitions and examples

An Arrow Debreu market $\mathsf{E} = \{X, \Omega_h, u_h, h = 1, \ldots, H\}$ has $H \geq 2$ traders, indexed $h = 1, \ldots, H$, $N \geq 2$ commodities and consumption or trading space[6] $X = R^N_+$ or $X = R^N$; elsewhere I have considered the case where instead of R^N one has a Hilbert space of infinite dimension.[7] The vector $\Omega_h \in R^N_+$ denotes trader h's property rights or initial endowment and $\Omega = (\sum_{h=1}^{H} \Omega_h)$ is the total endowment of the economy.[8] Traders may have zero endowments of some goods. Each trader h has a continuous and convex preference represented by $u_h : X \to R$. This paper treats in a unified way general convex preferences where simultaneously on every indifference surface for a given trader, the map $g(x) = Du(x)/\|Du(x)\|$ assigning to each vector the normalized utility gradient is either open or closed. Therefore either (i) all indifference surfaces contain no half lines (the map g is open) or (ii) the normalized gradients to any closed set of indifferent vectors define a closed set (the map g is closed). Some traders may have preferences of one type, and some of the other. Case (i) includes strictly convex preferences, and case (ii) linear preferences. All the assumptions and the results in this paper are ordinal,[9] therefore without loss of generality one normalizes utilities so that for all $h, u_h(0) = 0$ and $\sup_{\{x : x \in X\}} u_h(x) = \infty$. Preferences are increasing, i.e. $x > y \Rightarrow u_h(x) \geq u_h(y)$. When $X = R^N_+$ either indifference surfaces of positive utility are contained in the interior of X, R^N_{++}, such as Cobb–Douglas utilities, or if an indifference

[5] Chichilnisky and Heal [33] established domain restrictions which are necessary and sufficient for the existence of social choice rules which satisfy the axioms of [12, 13]: contractibility of the space of preferences is necessary and sufficient for the existence of social choice rules. Although these works deal with somewhat different axioms, they all find the same type of solution: a restriction of individuals' diversity.

[6] $R^N_+ = \{(x_1, \ldots, x_N) \in R^N : \forall i, x_i \geq 0\}$.

[7] Chichilnisky and Heal [31, 32] proved that limited arbitrage is necessary and sufficient for the existence of a competitive equilibrium in economies with infinitely many markets.

[8] If $x, y \in R^N$, $x \geq y \Leftrightarrow \forall i\, x_i \geq y_i$, $x > y \Leftrightarrow x \geq y$ and for some $i, x_i > y_i$, and $x \gg y \Leftrightarrow \forall i, x_i > y_i$.

[9] Namely independent of the utility representations.

surface of positive utility intersects a boundary ray,[10] it does so transversally.[11]

Definition 1. *A preference is uniformly non-satiated when it is represented by a utility u_h with a bounded rate of increase, e.g. for smooth preferences.*[12] $\exists \varepsilon, K > 0 : \forall x \in X, K > \| Du_h(x) \| > \varepsilon$.

Uniformly non-satiated preferences are rather common: for example, preferences represented by linear utilities are uniformly non-satiated.[13]

Proposition 2. *If a utility function $u_h : R^N \to R$ is uniformly non-satiated its indifference surfaces are within a uniform distance of each other, i.e. $\forall r, s \in R$, $\exists N(r,s) \in R$ such that $x \in u_h^{-1}(r) \Rightarrow \exists y \in u_h^{-1}(s)$ with $\| x - y \| \leq N(r,s)$.*

Proof. This is immediate from the definition. ∎

Assumption 1. When $X = R^N$, the preferences in the economy E are uniformly non-satiated.

The space of **feasible allocations** is $\Upsilon = \{(x_1, \dots, x_H) \in X^H : \sum_{h=1}^H x_h = \Omega\}$. The set of **supports to individually rational affordable efficient resource allocations** is

$$S(\mathsf{E}) = \{v \in R^N : \text{if } (x_1, \dots, x_H) \in \Upsilon \text{ with } u_h(x_h) \geq u_h(\Omega_h) \forall h = 1, \dots, H,$$
$$\langle v, x_h - \Omega_h \rangle = 0, \text{ and } \forall h\ u_h(z_h) \geq u_h(x_h) \text{ then } \langle v, z_h - x_h \rangle \geq 0\}. \tag{1}$$

The set of **prices orthogonal to the endowments** is[14]

$$\mathsf{N} = \{v \in R^N_+ - \{0\} : \exists h \text{ with } \langle v, \Omega_h \rangle = 0\}. \tag{2}$$

---

[10] A boundary ray r in R^N_+ is a set which consists of all the positive multiples of a vector $v \in \partial R^N_+$ i.e. $r = \{w \in R^N_+ : \exists \lambda > 0 \text{ s.t. } w = \lambda v\}$.

[11] This means that if $x \in \partial R^N_+$ and $u(x) > 0$, then the gradient $Du(x)$ is not orthogonal to ∂R^N_+ at λx, $\forall \lambda \geq 0$. This condition includes strictly convex preferences. Cobb–Douglas and CES preferences, many Leontieff preferences $u(x,y) = \min(ax, by)$, preferences which are indifferent to one or more commodities, such as $u(x,y,z) = \sqrt{x+y}$, preferences with indifference surfaces which contain rays of ∂R^N_+ such as $u(x,y,z) = x$, and preferences defined on a neighborhood of the positive orthant or the whole space, and which are increasing along the boundaries, e.g. $u(x,y,z) = x + y + z$.

[12] Smoothness is used to simplify notation only: uniform nonsatiation requires no smoothness. When preferences admit no smooth utility representation, then one requires $\exists \varepsilon, K > 0$: $\forall x, y, K\|x - y\| > |u(y) - u(x)|$ and $\sup_{\|x\| < \delta} |u(y + x) - u(x)| \geq \varepsilon\delta$.

[13] This includes linear preferences, piecewise linear preferences, Leontief preferences, preferences with indifference surfaces which intersect the boundary of the positive orthant (Arrow and Hahn 1971) and smooth utilities which are transversal to it.

[14] N is empty when $\forall h$, $\Omega_h \gg 0$.

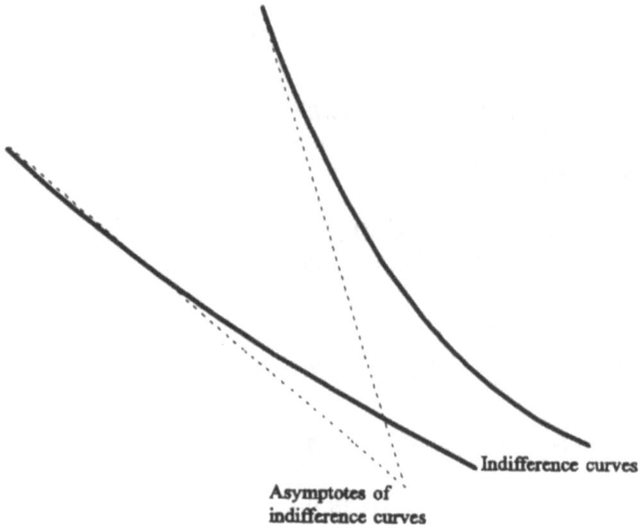

Fig. 1. This preference is not uniformly non-satiated because two indifference surfaces spread apart forever

The **utility possibility set** of the economy E is the set of feasible and individually rational utility allocations:

$$U(\mathsf{E}) = \{(V_1, \ldots, V_H) \colon \forall h, V_h = u_h(x_h) \geq u_h(\Omega_h) \geq 0, \text{ for some } (x_1, \ldots, x_H) \in \Upsilon\}.$$

The **Pareto frontier** of the economy E is the set of feasible, individually rational and efficient utility allocations:

$$P(\mathsf{E}) = \{U \in U(\mathsf{E}) \colon \sim \exists W \in U(\mathsf{E}) \colon W > U\}.$$

A **competitive equilibrium** of E consists of a price vector $p^* \in R^N_+$ and a feasible allocation $(x_1^*, \ldots, x_H^*) \in \Upsilon$ such that x_h^* optimizes u_h over the budget set $B_h(p^*) = \{x \in X \colon \langle x, p^* \rangle = \langle \Omega_h, p^* \rangle\}$.

2.1. Global and market cones

Two cases, $X = R^N$ and $X = R^N_+$, are considered separately.

- **Consider first** $X = R^N$.

Definition 3. *For trader h define the* **global cone** *of directions along which utility increases without bound:*

$$A_h(\Omega_h) = \{x \in X \colon \forall y \in X, \exists \lambda > 0 \colon u_h(\Omega_h + \lambda x) > u_h(y)\}$$

This cone was introduced in [18, 23, 24]; it contains global information on the economy and is new in the literature.[15] In ordinal terms, the rays of this cone intersect all indifference surfaces corresponding to bundles preferred by u_h to Ω_h. This cone and the part of its boundary along which utility never ceases to increase define:

$$G_h(\Omega_h) = \{x \in X : \sim \exists\, \text{Max}_{\lambda \geq 0} u_h(\Omega_h + \lambda x)\}$$

This cone is identical to the global cone defined in Chichilnisky (1995, p. 85, 4);[16] it treats all convex preferences in a unified way and under Assumption 1 it has a simple structure: when preferences have half lines in their indifferences $G_h(\Omega_h)$ equals $A_h(\Omega_h)$; when indifferences contain no half lines, then $G_h(\Omega_h)$ is its closure.

Definition 4. *The* ***market cone*** *of trader h is*

$$D_h(\Omega_h) = \{z \in X : \forall y \in G_h(\Omega_h),\ \langle z, y \rangle > 0\} \tag{3}$$

D_h is the cone of prices assigning strictly positive value to all directions of net trades leading to eventually increasing utility. This is a convex cone.

The following proposition establishes the structure of the global cones, and is used in proving the connection between limited arbitrage, equilibrium and the core:

Proposition 5. *If the function* $u_h : R^N \to R$ *is uniformly nonsatiated:* (i) *The interior of the global cone is*

$$G_h^o(\Omega_h) = A_h^o(\Omega_h)$$
$$= \{z \in G_h(\Omega_h) : \lim_{\lambda \to \infty} u_h(\Omega_h + \lambda z) = \infty\} \neq \emptyset.$$

(ii) *The boundary of the cone* $G_h(\Omega_h)$, $\partial G_h(\Omega_h)$, *contains* (a) *those directions along which utility increases towards a bounded value that is never reached:*

$$B_h(\Omega_h) = \{z \in \partial A_h(\Omega_h) : \forall \lambda > 0,\ u_h(\Omega_h + \lambda z) \neq \lim_{\lambda \to \infty} u_h(\Omega_h + \lambda z) < \infty\}$$

and (b) *those directions along which the utility eventually achieves a constant value:*

$$C_h(\Omega_h) = \{z \in \partial A_h(\Omega_h) : \exists N : \lambda, \mu > N \Rightarrow u_h(\Omega_h + \lambda z) = u_h(\Omega_h + \mu z)\},$$

---

[15] The global cone $A_h(\rho_h, \Omega_h)$ has points in common with Debreu's [34] "asymptotic cone" corresponding to the preferred set of u_h at the initial endowment Ω_h, in that along any of its rays utility increases. Under Assumption 1, its closure $\bar{A}_h(\Omega_h)$, equals the "recession" cone used by Rockafeller in 1970, but not generally: along the rays in $A_h(\Omega_h)$ utility increases beyond the utility level of any other vector in the space. This condition need not be satisfied by Debreu's asymptotic cones [35] or by "recession" cones. For example, for Leontief type preferences the recession cone through the endowment is the closure of the upper contour, which includes the indifference curve itself. By contrast, the global cone $A_h(\Omega_h)$ is the interior of the upper contour set. The global cone of a constant function is empty while the recession cone is the whole space. Related concepts appeared in Chichilnisky [10, 11]; otherwise there is no precedent in the literature for global cones.

[16] The cone $G_h(\Omega_h)$ is that in Chichilnisky [18] under Assumption 1, and it appears also in Monteiro Page and Wooders [51] which is a comment on Chichilnisky [18].

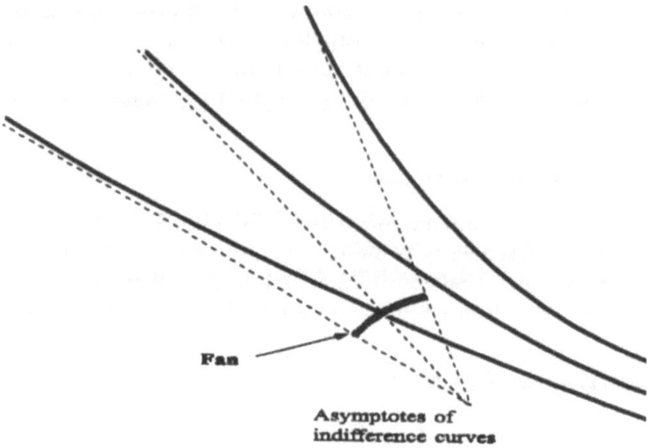

Fig. 2. This preference has a 'fan' of directions along which the utility values reach a bound. Assumption 1 is not satisfied. All the directions in the fan are in the recession cone but none in the global cone G_h or in the cone A_h.

(iii) *the interior of the global cone, its boundary and its closure and the cones* G_h *and* D_h *are uniform across all vectors in the space, i.e.* $\forall \Omega, \Lambda \in X$:

$$G_h^o(\Omega) = G_h^o(\Lambda) = A_h^o,$$

$$B_h(\Omega) \cup C_h(\Omega) \subset \partial G_h(\Omega) = \partial G_h(\Lambda) = \partial G_h,$$

$$G_h(\Omega) = G_h(\Lambda) = G_h, \; D_h(\Omega) = D_h(\Lambda) = D_h,$$

and in particular

$$\bar{G}_h(\Omega) = \bar{G}_h(\Lambda) = \bar{G}_h.$$

(iv) For general non-satiated preferences $G_h(\Omega_h)$ and $D_h(\Omega_h)$ may not be uniform.

Proof. See [23] and [26], Proposition 3, Appendix.

- **Consider next the case:** $X = R_+^N$.

Definition 6. *The* market cone of trader *h is*

$$D_h^+(\Omega_h) = D_h(\Omega_h) \cap S(\mathsf{E}) \;\; \text{if } S(\mathsf{E}) \subset \mathsf{N},$$
$$= D_h(\Omega_h) \qquad \text{otherwise.} \tag{4}$$

where $S(\mathsf{E})$ *and* N *are defined in* (1) *and* (2).[17]

[17] The market cone D_h^+ is the whole consumption set $X = R^{N+}$ when $S(E)$ has a vector assigning strictly positive income to all individuals. If some trader has zero income, then this trader must have a boundary endowment.

There is no analog to Proposition 5 when $X = R^N_+$; indeed, when $X = R^N_+$ the market cones $D_h^+(\Omega_h)$ typically vary with the initial endowments.

Proposition 7. *When $X = R^N_+$ and an indifference surface of u_h corresponding to a positive consumption bundle $x > 0$ intersects a boundary ray $r \subset \partial X$, then $r \in G_h(0)$.*

Proof. Recall that we assumed $u_h(0) = 0$, and that the preference's indifference surfaces of positive utility are either (a) contained in the interior of R^N_+, R^N_{++}, or (b) they intersect a boundary ray r of R^N_+ and do so transversally. In case (a) the proposition is satisfied trivially, because no indifference surface of strictly positive value ever intersects the boundary of R^N_+. In case (b) the proposition follows immediately from the definition of transversality. Observe that it is possible that $sup_{x \in r}(u_h(x)) < \infty$. ■

Definition 8. *The* core *of the economy E is the set of allocations which no coalition can improve upon within its own endowments:*

$$C(E) = \{(x_1, \dots, x_H) \in R^{N \times H} : \sum_h (x_h - \Omega_h) = 0 \text{ and } \sim J \subset \{1, \dots, H\}:$$

$$\times \sum_{j \in J}(y_j - \Omega_j) = 0, \forall_j \in J, u_j(y_j) \geq u_j(x_j),$$

$$\text{and } \exists_j \in J : u_j(y_j) > u_j(x_j)\}.$$

3. Limited arbitrage: definition and examples

This section provides the definition of limited arbitrage. It gives an intuitive interpretation for limited arbitrage in terms of gains from trade, and contrasts limited arbitrage with the arbitrage concept used in financial markets. It provides examples of economies with and without limited arbitrage.

Definition 9. *When $X = R^N$,* E *satisfies* ***limited arbitrage*** *if*

$$(LA) \bigcap_{h=1}^{H} D_h \neq \emptyset.$$

Definition 10. *When $X = R^N_+$,* E *satisfies* ***limited arbitrage*** *if*

$$(LA^+) \bigcap_{h=1}^{H} D_h^+(\Omega_h) \neq \emptyset. \tag{5}$$

3.1. Interpretation of limited arbitrage as bounded gains from trade when $X = R^N$

Gains from trade are defined by

$$G(E) = \sup\left\{\sum_{h=1}^{H} u_h(x_h) - u_h(\Omega_h)\right\}, \text{ where}$$

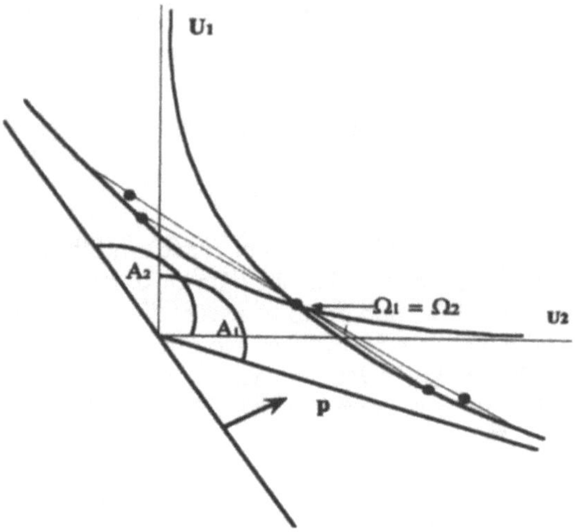

Fig. 3. Limited arbitrage is satisfied: feasible allocations lead to bounded utility increases

$$\sum_{h=1}^{H} (x_h - \Omega_h) = 0, \quad \text{and} \quad \forall h, \quad u_h(x_h) \geq u_h(\Omega_h) \geq 0.$$

The proposition below applies to preferences where the normals to a closed set of indifferent vectors defines a closed set, case (ii); its corollary following it applies both to case (i) and (ii):

Proposition 11. *In case* (ii), *the economy* E *satisfies limited arbitrage if and only if gains from trade are bounded i.e. if and only if*[18]

$$G(\mathrm{E}) < \infty.$$

For a proof see e.g. [23] and Theorem 1 proved in the Appendix of [26]; its sufficiency part is valid for all preferences satisfying Assumption 1, case (i) or case (ii), so that:

Corollary 12. *For all economies with uniformly non-satiated preferences, limited arbitrage implies bounded gains from trade.*[19]

[18] The expression $G(\mathrm{E}) < \infty$ holds when $\forall h$, $\sup_{\{x:x \in X\}} u_h(x) = \infty$ as considered here; it must be replaced by $G(\mathrm{E}) < \sup_{\{x:x \in X^H: \sum x_h = \Omega, u_h(x_h) \geq u_h(\Omega_h)\}} (\sum_{h=1}^{H} u_h(x_h) - u_h(\Omega_h)) - k$, for some positive k, when $\forall h \sup_{\{x:x \in X\}} u_h(x) < \infty$.

[19] The simplest illustration of the link between limited arbitrage and no-arbitrage is an economy E where the traders' initial endowments are zero, $\Omega_h = 0$ for $h = 1, 2$, and the set of gradients to indifference surfaces are closed. Here no-arbitrage at the initial endowments means that there are no trades which could increase the traders' utilities at zero cost: gains from trade in E are zero. By contrast, E has limited arbitrage when no trader can increase utility beyond a given bound at zero cost: gains from trade are bounded.

In brief: *no-aribtrage* requires that there should be no gains from trade at zero cost while *limited arbitrage* requires that there should be only bounded utility arbitrage or limited gains from trade. In linear economies, limited arbitrage "collapses" into no-arbitrage.

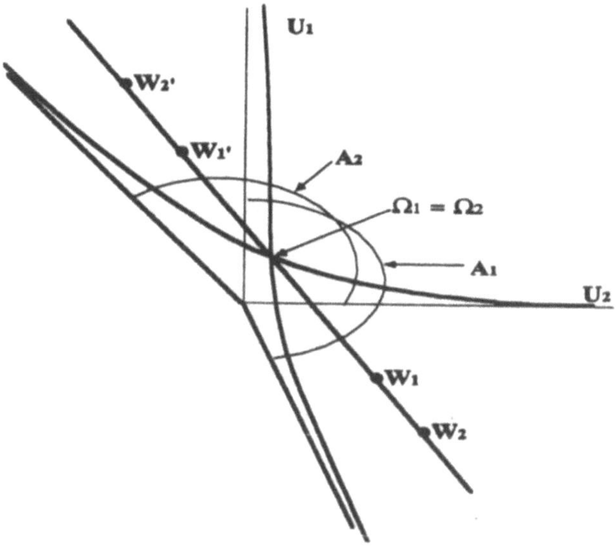

Fig. 4. Limited arbitrage is not satisfied: there exist a feasible unbounded sequence of allocations, (W_1, W'_1), (W_2, W'_2), ..., along which both traders' utility never ceases to increase

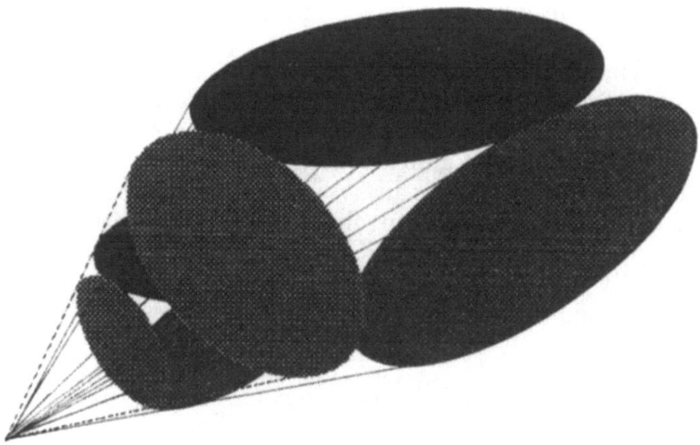

Fig. 5. Three traders in R^3. Every two traders's subeconomy has limited arbitrage but the whole economy does not

3.2. Examples of markets with and without limited arbitrage

Example 13. Figures 3 and 4 above illustrate an economy with two traders trading in $X = R^2$; in Fig. 3 the market cones intersect and the economy has limited arbitrage. In Fig. 4 the market cones do not intersect and the economy does not have limited arbitrage. Figure 5 below illustrates three traders trading in $X = R^3$; each two market cones intersect, but the three market

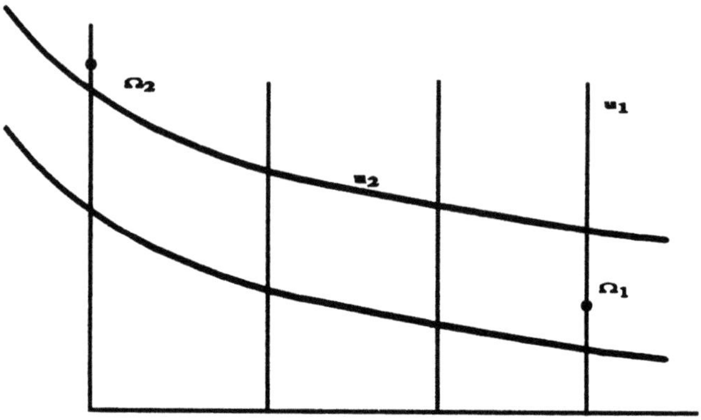

Fig. 6. Limited arbitrage fails. Trader two only owns one good, to which the first trader is indifferent

cones do not intersect, and the economy violates limited arbitrage. This figure illustrates the fact that the union of the market cones may fail to be contractible: indeed, this failure corresponds to the failure of the market cones to intersect, as proven in Chichilnisky [15] and [19].

Example 14. When the consumption set is $X = R^N_+$, limited arbitrage is always satisfied if all indifference surfaces through positive consumption bundles are contained in the interior of X, R^N_{++}. Examples of such preferences are those given by Cobb–Douglas utilities, or by utilities with constant elasticity of substitution (CES) with elasticity of substitution $\sigma < 1$. This is because all such preferences have as global cone the positive orthant (or its closure), and therefore their market cones always interesect. These preferences are very similar to each other on choices involving large utility levels: this is a form of similarity of preferences. Economies where the individuals' initial endowments are strictly interior to the consumption set X always satisfy the limited arbitrage condition in the case $X = R^N_+$, since in this case $\forall h$, $R^N_{++} \subset D^+_h(\Omega_h)$ for all $h = 1, \dots, H$.

Example 15. When $X = R^N_+$ the limited arbitrage condition may fail to be satisfied when some trader's endowment vector Ω_h is in the boundary of the consumption space, ∂R^N_+, and at all supporting prices in $S(\mathsf{E})$ some trader has zero income, i.e. when $\forall p \in S(\mathsf{E})$ $\exists h$ such that $\langle p, \Omega_h \rangle = 0$. In this case, $S(\mathsf{E}) \subset \mathsf{N}$. See e.g. [23, 24, 26]. This case is illustrated in Figure 6 above; it is a rather general case which may occur in economies with many individuals and with many commodities. When all individuals have positive income at some price $p \in S(\mathsf{E})$, then limited arbitrage is always satisfied since by definition in this case $\forall h, = R^N_{++} \subset D^+_h(\Omega_i)$ for all $h = 1, \dots, H$.

Example 16. A competitive equilibrium may exist even when some traders have zero income, showing that Arrow's "resource relatedness" condition [3] is sufficient but not necessary for existence of an equilibrium. Figure 7 below illustrates an economy where at all supporting prices some trader has zero income: $\forall p \in S(\mathsf{E})$ $\exists h$ such that $\langle p, \Omega_h \rangle = 0$, i.e. $S(\mathsf{E}) \subset \mathsf{N}$; in this economy,

Fig. 7. Equilibrium exists even when one trader has zero income

however, limited arbitrage is satisfied so that a competitive equilibrium exists. The initial allocation and a price vector assigning value zero to the second good defines such an equilibrium.

4. Limited arbitrage and the compactness of the Pareto frontier

The Pareto frontier $P(\mathsf{E})$ is the set of feasible, efficient and individually rational utility allocations. With H traders it is a subset of R^H_+. Proving the boundedness and closedness of the Pareto frontier is a crucial step in establishing the existence of a competitive equilibrium and the non-emptiness of the core. The main theorem of this section shows that limited arbitrage is necessary and sufficient for this; it appeared first in [16, 18, 23, 24, 26].

There is a novel feature of the results which are presented here, a feature which is shared which those that were previously established in Chichilnisky [18, 20, 23, 24, 26], and Chichilnisky and Heal [30, 31]. It starts from the observation that the compactness of the Pareto frontier need not imply the compactness of the set of feasible commodity allocations. Indeed, such boundedness is not used in this paper, nor was it used in the results of Chichilnisky [18, 20, 23, 24, 26] and Chichilnisky and Heal [31]: these are the first results in the literature proving the existence of equilibrium and the non-emptiness of the core in economies where limited arbitrage holds and the set of feasible and individually rational allocations is generally unbounded.

Example 17. Figure 8 shows that the Pareto frontier may fail to be closed even in finite dimensional models, provided the consumption set is the whole Euclidean space. It shows two traders with indifference curves having the line $y = -x$ as asymptote. Consumption sets are the whole space and feasible allocations are those which sum to zero. Utility functions are $u_i = x_i + y_i \pm e^{-(x_i - y_i)}$, $i = 1, 2$. Limited arbitrage rules out such cases. Another example is a two-agent economy where both agents have linear preferences: if these preferences are different the set of feasible utility allocations is unbounded. Of course, limited arbitrage rules out such situations.

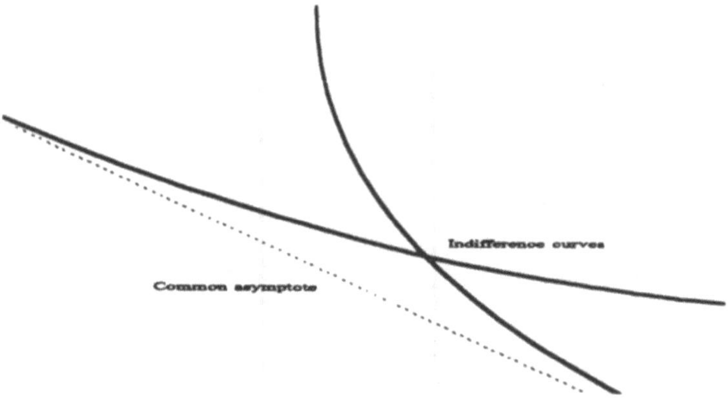

Fig. 8. The Pareto frontier may fail to be closed even in finite dimensions

Theorem 18. *Consider an economy* E *as defined in Sect.* 1. *Then* ***limited arbitrage*** *is necessary and sufficient for the compactness of the Pareto frontier.*

See [16, 18, 23, 24] and the Proof of Theorem 1 in the Appendix of [26].

5. Competitive equilibrium and limited arbitrage

This section links the existence of a competitive equilibrium with limited arbitrage. The result is that limited arbitrage is simultaneously necessary and sufficient for the existence of a competitive equilibrium.

The result presented below was established first in Chichilnisky [16, 18, 20, 21, 23, 24, 26, 30] for preferences which are either all in case (i) e.g. strictly convex, or in case (ii), e.g. they have indifference surfaces with a closed set of gradient directions. The result presented here extends these earlier results in that it deals in a unified way with non-satiated convex preferences; in the same economy there may be a mixture of preferences of type (i) and (ii);[20] it appeared first in [20, 23].

The condition of limited arbitrage need not be tested on all traders simultaneously: in the case of R^N, it needs only be satisfied on subeconomies

[20] The results on equilibrium in this paper originated from a theorem in Chichilnisky and Heal [30], a paper which was submitted for publication in 1984, nine years before it appeared in print: these dates are recorded in the printed version [30]. Chichilnisky and Heal [30, 31] provide a no-arbitrage condition and prove that it is sufficient for the existence of a competitive equilibrium and the compactness of the utility set in Arrow Debreu economies with or without short sales, with infinitely or finitely many markets and with general preferences which are convex or strictly convex. Subsequent to [30] Werner [57] proved existence results for markets with short sales, relying on a no-arbitrage condition defined from recession cones. For Hart-type models which are incomplete as they lack futures markets and which do not have the generality of the Arrow Debreu model., Page [54] uses Werner's no-arbitrage condition to prove existence of a market equilibrium. The equilibria in Hart-type models are generally inefficient, and their allocations are not in the Pareto frontier.

with no more traders than the number of commodities in the economy, N, plus one.

Definition 19. *A k-trader sub-economy of* E *is an economy F consisting of a subset of* $k \leq H$ *traders in* E, *each with the endowments and preferences as in* E: $\mathsf{F} = \{X, u_h, \Omega_h, h \in J \subset \{1, \ldots, H\}$, *cardinality* $(J) = k \leq H\}$.

Theorem 20. *The following four properties of an economy* E *with trading space* R^N *are equivalent:*
(i) E *has a competitive equilibrium*
(ii) *Every sub economy of* E *with at most* $N+1$ *traders has a competitive equilibrium*
(iii) E *has limited arbitrage*
(iv) E *has limited arbitrage for any subset of traders with no more that* $N+1$ *members.*

Proof. See [20, 23] and the proof of Theorem 2 in the Appendix of [26] for (i)$\Leftrightarrow$(iii) and (ii)$\Leftrightarrow$(iv). That (iii)$\Leftrightarrow$(iv) follows from Helley's theorem, which is a corollary of the following theorem established in Chichilnisky [19]: Consider a family $\{U_i\}_{i=1,\ldots,H}$ of convex sets in R^N, $H, N \geq 1$. Then

$$\bigcap_{i=1}^{H} U_i \neq \emptyset \quad \text{if and only if} \bigcap_{j \in J} U_i \neq \emptyset$$

for any subset of indices $J \subset \{1, \ldots, H\}$ having at most $N+1$ elements.

In particular, an economy E as defined in Section 2 satisfies limited arbitrage, if and only if it satisfies limited arbitrage for any subset of $k = N+1$ traders, where N is the number of commodities in the economy E. ■

6. Limited arbitrage, equilibrium and the core

The following is a corollary of Theorem 20; it was first established in [20, 21, 23, 28]:

Theorem 21. *Consider an economy* $\mathsf{E} = \{X, u_h, \Omega_h, h = 1, \ldots, H\}$, *where* $H \geq 2$, $X = R^N$ *and* $N \geq 1$, *or X is a Hilbert space H. Then the following four properties are equivalent:*

(*i*) *The economy* E *has limited arbitrage;*
(*ii*) *The economy* E *has a core;*
(*iii*) *The economy* E *has a competitive equilibrium;*
(*iv*) *When* $X = R^N$, *every subeconomy of* E *with at most* $N+1$ *traders has a competitive equilibrium and a core.*

Proof. For a proof of (i)$\Leftrightarrow$(ii)$\Leftrightarrow$(iii) see e.g. [28], which contains also a discussion of the literature.[21]

---

[21] This result was presented at the January 3–5, 1994 Annual Meetings of the Econometric Society in Boston. Page acted as referee for [24] during 1993 and attended this presentation. Subsequent to my January 1994 presentation in Boston I received a working paper by Page and Wooders on a related topic; see Monteiro Page and Wooders [51] which is a comment on Chichilnisky [19, 22] and [24], and the response in Chichilnisky [26].

Finally I establish (iv)$\Leftrightarrow$(i): this follows directly from the fact that an economy has limited arbitrage if and only if every subeconomy of at most $N + 1$ traders has limited arbitrage, Theorem 36 in the Appendix. ■

7. Limited arbitrage and social choice

Limited arbitrage is also crucial for achieving resource allocation via social choice. Two main approaches to social choice are studied here. One is Arrow's: his axioms of social choice require that the social choice rule Φ be non-dictatorial, independent of irrelevant alternatives, and satisfy a Pareto condition Arrow [2]. A second approach requires, instead, that the rule Φ be continuous, anonymous, and respect unanimity, Chichilnisky [12, 13]. Both approaches have led to corresponding impossibility results [2, 12, 13]. Though the two sets of axioms are quite different, it has been shown recently that the impossibility results which emerge from them are in a sense equivalent, see Baryshnikov [7]. Furthermore, as is shown below, limited arbitrage is closely connected with both sets of axioms. Economies which satisfy limited arbitrage admit social choice rules with either set of axioms. Therefore, in a well defined sense, the social choice problem can only be solved in those economies which satisfy limited arbitrage.

How do we allocate resources by social choice? Social choice rules assign a social preference $\Phi(u_1, \dots, u_H)$ to each list $(u_1, \dots, u_H)$ of individual preferences of an economy E.[22] The social preference ranks allocations in $R^{N \times H}$, and allows one to select an optimal feasible allocation. This is the allocation obtained via social choice.

The procedure requires, of course, that a social choice rule Φ exists: the role of limited arbitrage is important because it ensures existence. This will be established below. I prove here that limited arbitrage is necessary and sufficient for resolving Arrow's paradox, when the domain of individual preferences are those in the economy, and the choices are those feasible allocations which give large utility value.[23]

Limited arbitrage provides a restriction on the relationship between individual preferences under which social choice rules exist. A brief background on the matter of preference diversity follows.

Arrow's impossibility theorem established that in general a social choice rule Φ does not exist: the problem of social choice has no solution unless individual preferences are restricted. Duncan Black [8] established that the "single peakedness' of preferences is a sufficient restriction to obtain majority rules. Using different axioms, Chichilnisky [12, 13] established also that a social choice rule Φ does not generally exist; subsequently Chichilnisky and Heal [30] established a necessary and sufficient restriction for the resolution of the social choice paradox: the *contractibility* of the space of preferences.

---

[22] In the economy E the traders' preferences are defined over private consumption $u_i: R^N \to R$, but they define automatically preferences over allocations in $R^{N \times H}$: $u_i(x_1, \dots, x_H) \geq u_i(y_1, \dots, y_H) \Leftrightarrow u_i(x_i) \geq u_i(y_i)$.

[23] As defined below the concept of "large utility values" is purely ordinal; it is defined relative to the maximum utility value achieved by a utility representation.

Contractibility can be interpreted as a limitation on preference diversity, Heal [44]. In all cases, therefore, the problem of social choice is resolved by restricting the diversity of individual preferences. The main result in this section is that the restriction on individual preferences required to solve the problem is precisely limited arbitrage. The connection between limited arbitrage and contractibility is discussed below.

The section is organized as follows. First I show in Proposition 24 that the economy E satisfies limited arbitrage if and only if it contains no *Condorcet cycles* on choices of large utility values. Condorcet cycles are the building blocks of Arrow's impossibility theorem, and are at the root of the social choice problem. On the basis of Proposition 24, I prove in Theorem 28 that limited arbitrage is necessary and sufficient for resolving Arrow's paradox on allocations of large utility values.

Definition 22. *A Condorcet cycle is a collection of three preferences over a choice set X, represented by three utilities $u_i\colon X \to R$, $i = 1, 2, 3$, and three choices α, β, γ within a feasible set $Y \subset X$ such that $u_1(\alpha) > u_1(\beta) > u_1(\gamma)$, $u_2(\gamma) > u_2(\alpha) > u_2(\beta)$ and $u_3(\beta) > u_3(\gamma) > u_3(\alpha)$.*

In an economy with finite resources $\Omega \geq 0$, the social choice problem is about the choice between allocations of these resources. Choices are in $X = R^{N \times H}$. An allocation $(x_1, \ldots, x_H) \in R^{N \times H}$ is *feasible* if $\sum_i x_i - \Omega = 0$. Consider an economy E as defined in Section 2. Preferences over private consumption are increasing, $u_h(x) > u_h(y)$ if $x \gg y \in R^N$, utilities are uniformly non-satiated (Assumption 1), and indifference surfaces which are not bounded below have a closed set of gradients, so that $G_h = A_h$. While the preferences in E are defined over private consumption, they naturally define preferences over allocations, as follows: define $u_h(x_1, \ldots, x_H) \geq u_h(y_1, \ldots, y_H) \Leftrightarrow u_h(x_h) \geq u_h(y_h)$. Thus the preferences in the economy E induce naturally preferences over the feasible allocations in E.

Definition 23. *The family of preferences $\{u_1, \ldots, u_H\}$, $u_h\colon R^N \to R$ of an economy E has a Condorcet cycle of size k if for every three preferences $u_1^k, u_2^k, u_3^k \in \{u_1, \ldots, u_H\}$ there exists three feasible allocations $\alpha^k = (\alpha_1^k, \ldots, \alpha_H^k) \in X^H \subset R^{3 \times N \times H}$; $\beta^k = (\beta_1^k, \ldots, \beta_H^k)$ and $\gamma^k = (\gamma_1^k, \ldots, \gamma_H^k)$ which define a Condorcet cycle, and such that each trader $h = 1, \ldots, H$, achieves at least a utility level k at each choice:*

$$\min_{h=1,\ldots,H} \{[u_h^k(\alpha_h^k), u_h^k(\beta_h^k), u_h(\gamma_h^k)]\} > k.$$

The following shows that limited arbitrage eliminates Condorcet cycles on matters of great importance, namely on those with utility level approaching the supremum of the utilities, which for simplicity and without loss of generality we have assumed to be ∞. Preferences are in case (ii):

Proposition 24. *Let E be a market economy with short sales ($X = R^N$) and $H \geq 3$ traders. Then E has social diversity if and only if its traders' preferences have Condorcet cycles of every size. Equivalently, E has limited arbitrage if and only for some $k > 0$, the traders' preferences have no Condorcet cycles of size larger than k.*

Proof. Consider an economy with Condorcet cycles of all sizes. For each $k > 0$, there exists three allocations denoted $(\alpha^k, \beta^k, \gamma^k) \in R^{3 \times N \times H}$ and every three traders which define a Condorcet triple of size k. By definition, for every k, each of the three allocations is feasible, for example, with $\Omega = 0$ $\alpha^k = (\alpha_1^k, \ldots, \alpha_H^k) \in R^{N \times H}$, and $\sum_{i=1}^{H} (\alpha_i^k) = 0$. Furthermore $\min_{h=1,\ldots,H} \{[u_h^k(\alpha_h^k), u_h^k(\beta_h^k), u_h(\gamma_h^k)]\} > k$, so that e.g. $\lim_{k \to \infty} (u_h(\alpha_h^k)) = \infty$. There exist therefore a sequence of allocations $(\theta^k)_{k=1,2\ldots} = (\theta_1^k, \ldots, \theta_H^k)_{k=1,2\ldots}$ such that $\forall k$, $\sum_{h=1}^{H} \theta_h^k = 0$ and $\forall h \sup_{k \to \infty}(\inf_h u_h(\theta_h^k)) = \infty$. This implies that E has unbounded gains from trade, which contradicts Proposition 11. Therefore E cannot have Condorcet cycles of every size.

Conversely, if E has no limited arbitrage, for any $k > 0$, there exist a feasible allocation $(a_1^k, a_2^k, \ldots, a_H^k)$, such that $\sum_{h=1}^{H} a_h^k \leq 0$, and $\forall h$, $u_h(a_h^k) \geq k$. For each integer $k > 0$, and for a small enough $\epsilon > 0$ define now the vector $\Delta = (\varepsilon, \ldots, \varepsilon) \in R_+^N$ and the following three allocations: $\alpha^k = (ka_1^k, ka_2^k - 2\Delta, ka_3^k + 2\Delta, ka_4^k, \ldots, ka_H^k)$, $\beta^k = (ka_1^k - \Delta, ka_2^k, ka_3^k + \Delta, ka_4^k, \ldots, ka_H^k)$, and $\gamma^k = (ka_1^k - 2\Delta, ka_2^k - \Delta, ka_3^k + 3\Delta, ka_4^k, \ldots, ka_H^k)$. Each allocation is feasible, e.g. $ka_1^k + ka_2^k - 2\Delta + ka_3^k + 2\Delta + ka_4^k + \ldots + ka_H^k = k(\sum_{h=1}^{H} a_h^k) \leq 0$. Furthermore for each $k > 0$ sufficiently large, the three allocations $\alpha^k, \beta^k, \gamma^k$ and the traders $h = 1, 2, 3$, define a Condorcet cycle of size k: all traders except for 1, 2, 3, are indifferent between the three allocations and they reach a utility value at least k, while trader 1 prefers α^k to β^k to γ^k, trader 3 prefers γ^k to α^k to β^k, and trader 2 prefers β^k to γ^k to α^k. Observe that this construction can be made for any three traders within the set $\{1, 2, \ldots, H\}$. This completes the proof. ■

The next result uses Proposition 24 to establish the connection between limited arbitrage and Arrow's theorem. Consider Arrow's three axioms: Pareto, independence of irrelevant alternatives, and non-dictatorship. The social choice problem is to find a social choice rule $\Phi: P^j \to P$ from individual to social preferences satisfying Arrow's three axioms; the domain for the rule Φ are profiles of individual preferences over allocations of the economy E. Recall that each preference in the economy E defines a preference over feasible allocations in E.

Definition 25. *The economy* E *admits a resolution of Arrow's paradox if for any number of voters $j \geq 3$ there exists a social choice function from the space $P = \{u_1, \ldots, u_H\}$ of preferences of the economy* E *into the space Q of complete transitive preference defined on the space of feasible allocations of* E, $\Phi: P^j \to Q$, *satisfying Arrow's three axioms.*

Definition 26. *A feasible allocation $(\alpha_1^k, \ldots, \alpha_H^k) \in R^{N \times H}$ has utility value k, or simply value k, if each trader achieves at least a utility level k:*

$$\min_{h \in H} \{[u_1^k(\alpha_1^k), \ldots, u_H(\alpha_H^k)]\} > k.$$

Definition 27. *Arrow's paradox is said to be resolved on choices of large utility value in the economy* E *when for all $j \geq 3$ there exists a social choice function $\Phi: P^j \to Q$ and a $k > 0$ such that Φ is defined on all profiles of j preferences in* E, *and it satisfies Arrow's three axioms when restricted to allocations of utility value exceeding k.*

Theorem 28. *Limited arbitrage is necessary and sufficient for a resolution of Arrow's paradox on choices of large utility value in the economy* E.

Proof. Necessity follows from Proposition 24, since by Arrow's axiom of independence of irrelevant alternatives, the existence of one Condorcet triple of size k suffices to produce Arrow's impossibility theorem on feasible choices of value k in our domain of preferences, see Arrow (1951). Sufficiency is immediate: limited arbitrage eliminates feasible allocation of large utility value by Proposition 11, because it bounds gains from trade. Therefore it resolves Arrow's paradox, because this is automatically resolved in an empty domain of choices. ■

7.1. Anonymity and respect of unanimity

Consider now the second approach to social choice, Chichilnisky [12, 13] which seeks continuous anonymous social choice rules which respect unanimity. The link connecting arbitrage with social choices is still very close but it takes a different form. In this case the connection is between the contractibility of the space of preferences, which is necessary and sufficient for the existence of continuous, anonymous rules which respect unanimity (Chichilnisky and Heal [33]) and limited arbitrage.

Continuity is defined in a standard manner; anonymity means that the social preference does not depend on the order of voting. Respect of unanimity means that if all individuals have identical preferences overall, the social preference is this common preference; it is a very weak version of the Pareto condition. It was shown in Chichilnisky [12, 13] that, for general spaces of preferences, there exist no social choice rules satisfying these three axioms. Subsequently Chichilnisky and Heal [33] established that contractibility of the space of preferences is exactly what is needed for the existence of social choice rules.

Formally: consider a general topological space χ consisting of preferences over the space of allocations X^K for $K \geq 2$ individuals. The preferences in χ need not be those of a market economy. A *K-profile of individual preferences* is a list of preferences for the K individuals, i.e. a K-tuple of preferences in the space χ, denoted $(\kappa_1 \ldots \kappa_K) \in \chi^K$. The *social choice problem* is defined on any space of preferences χ as the problem of finding for all $K \geq 2$ a map $\phi : \chi^K \to \chi$ such that

(*A1*) ϕ is *continuous*

(*A2*) ϕ is *symmetric or anonymous*, i.e. $\phi(\kappa_1, \ldots, \kappa_K) = \phi(\pi(\kappa_1), \ldots, \pi(\kappa_K))$ for any permutation π of the set $\{1, \ldots, K\}$, and

(*A3*) ϕ *respects unanimity*, i.e. $\phi(\kappa_1, \ldots, \kappa_K) = \kappa$ if $\kappa_i = \kappa_j$ for all $i, j \in \{1, \ldots, K\}$.

Axioms (A1)–(A3) were introduced in Chichilnisky [12, 13]. A comparison between these axioms and Arrow's classic axioms of social choice is discussed in Section 13. It was proven in Theorem 1 of Chichilnisky [12] that these three axioms are generally inconsistent, in the sense that when χ is *the space of all non-trivial smooth preferences* defined on Euclidean choice spaces, *there exists no map ϕ satisfying these three axioms.* The result is valid whether or not the preferences admit satiation. However, when respect of unanimity (A3) is replaced by a Pareto condition, the impossibility result of Chichilnisky [12] holds even when the total indifference between all choices is allowed as the social preference.[24]

---

[24] The *Pareto condition* requires that when all preferences in a profile prefer one choice y to another x, so does the social preference.

It is worth observing that *the following result is valid for any topology on the space of preferences T*. In this sense this result is analogous to a fixed point theorem or to a maximization theorem: whatever the topology, a continuous function from a compact convex space to itself has a fixed point and a continuous function of a compact set has a maximum. All these statements, and the one below, apply independently of the topology chosen, for general parafinite CW complexes:

Theorem 29. *Let T be a connected space of preferences endowed with any topology. Then T admits a continuous anonymous map Φ respecting unanimity*

$$\Phi : T^k \to T$$

for every $k \geq 2$, if and only if T is contractible.

Proof. See Chichilnisky and Heal [33]. ■

The close relation between contractibility and non-empty intersection – which is limited arbitrage – follows from the following theorem:

Theorem 30. *Let $\{U_i\}_{i=1 \ldots I}$ be a family of convex sets in R^N. The family has a non-empty intersection if and only if every subfamily has a contractible union:*

$$\bigcap_{i=1}^{I} U_i \neq \phi \quad \Leftrightarrow \quad \bigcup_{i \in J} U_i \text{ is contractible } \forall J \subset \{1, \ldots, I\}.$$

Proof. See Chichilnisky [15, 19]. ■

This theorem holds for general excisive families of sets, including acyclic families and even *simple families* which consist of sets which need not be convex, acyclic, open or even connected. This theorem was shown to imply Helley's theorem, the Knaster Kuratowski Marzukiewicz theorem and Brouwer's fixed point theorem [19], but it is not implied by them.

7.1.1. Social allocations

Consider any space χ of preferences over the space of allocations X^r; there are $r \geq 2$ individuals, each with a preference $\kappa_i \in \chi$, $i = 1, \ldots, r$. The preferences in χ may or may not be part of a market economy. When a social choice map $\phi : \chi^r \to \chi$ exists satisfying the three axioms: continuity (A1), anonymity (A2), and respect of unanimity (A3) it defines *a social allocation for the space of preferences* χ as follows: For each profile of individual preferences $(\kappa_1, \ldots, \kappa_r) \in \chi^r$ a *social allocation* is a resource allocation in χ^r which is optimal within the set of feasible allocations Υ according to the social preference $\theta(\kappa_1, \ldots, \kappa_r) \in \chi$. Such a resource allocation is located by a social choice rule which satisfies the three ethical axioms (A1)–(A3).

8. Similarity of preferences

We aim to define precisely what is meant by preferences being similar, and to show how to define populations of individuals having preferences similar to those in the market E. The concepts of *similarity and diversity* are defined from the preferences and endowments in the economy E. An essential characteristic of the agents is their global cones. A geometric interpretation is that a preference σ is *similar* to another preference ρ when σ increases along those directions which give ρ unbounded utility gains, and only along those directions. In particular, a preference σ is dissimilar to another ρ if – and only if – there exist a direction along which ρ can achieve unbounded increases in utility, yet along the same direction, σ's utility decreases. Intuitively the preferences ρ and σ should not be considered similar in this latter case. We therefore consider a universe of preferences in which individuals' preferred directions are similar to those of the individuals in the market in the sense that their gradients have positive inner products with the global cones of some trader in the economy E.

Formally, consider an economy $E = \{X, \Omega_i, \rho_i, i = 1, \ldots, H\}$. Each preference ρ_i in E is defined over i's private consumption set X. However, as we saw between definitions 22 and 23 above, ρ_i also defines a preference over all allocations in X^r for any number of individuals $r \geq 1$: an allocation $(x_1, \ldots, x_r) \in X^r$ is preferred by individual i in position j to another $(y_1, \ldots, y_r)$ if and only if ρ_i prefers x_j to y_j. We may therefore consider ρ_i as a preference defined over allocations, using for it the same notation when the meaning is clear from the context.

Consider first the case $X = R^N$. Let η be any concave, smooth preference defined over allocations in X^r, $r \geq 1$. Note that, in general, a preference over allocations may not be monotonic. We seek to formalize the notion that η is *similar* to the preference of trader i in the market E if it agrees with the preference ρ_i of i on important choices: the preference η must increase in the directions of individual i's global cone. This is formalized in the following definition of *similarity to the preference of trader i.*

Let the normal to the indifference surface of the preference η at the allocation $\xi \in X^r$ be denoted $G\eta(\xi)$; it is an $N \times r$ vector indicating the direction of increase of preference η at the social allocation $\xi \in R^r$. Recall that η is a preference over allocations. This normal always exists and is unique because we assumed that the preference η is smooth:

Definition 31. When $X = R^N$ an individual preference η over resource allocations in X^r is said to be similar to the preference of trader i of the economy $E = \{X, \rho_i, \Omega_i, i = 1, \ldots, H\}$ in position j when

$$\forall \xi \in X^r \quad G\eta^j(\xi) \in D_i(\rho_i, \Omega_i), \tag{6}$$

where $D_i(\rho_i, \Omega_i)$ is the market cone of individual i. This means that the gradient of η has a strictly positive inner product with the vectors of the global cone $A_i(\rho_i, \Omega_i)$ of individual i in position j, as defined in Section 2.

8.1. Private and public preferences

A preference η is *private* when it is indifferent to the consumption of anyone else, and it is *public* otherwise. Therefore a preference ρ_i is *private* when it is defined over individual i's (private) consumption set X; it is public if defined over allocations, X^H. For any number $r \geq 2$ of individuals, a private preference ρ_i defines a *public preference*: $(x_1, \ldots, x_r) \in X^r$ is preferred by individual i in position $j \in \{1, \ldots, r\}$ to $(y_1, \ldots, y_r) \in X^r \Leftrightarrow x_j \succeq_{\rho_i} y_j$. When a preference is private and the individual occupies position $j \in \{1, \ldots, r\}$, then its normal

$G\eta(\xi) \in R^{r\times N}$ has only N non-zero components, those in position j, indicated $G\eta^j(\xi)$. A private preference is therefore similar to that of an individual i in the economy E in one position j. By contrast, if a preference is not private, it may in principle be similar to those of different individuals i in different positions $j = 1, \dots, r$. The result of this paper apply equally well when we consider solely private individual preferences, or when individual preferences are either private or public. When the space of preferences consists solely of private preferences, the concept of respect of unanimity must be modified slightly.[25] We consider here spaces consisting of preferences which are either private or public.

We now define *similarity of a preference* with respect to a set K of traders in the economy E (as opposed to similarity to the preference of a trader i in position $j \in \{1, 2, \dots, r\}$, which was defined in (6)):

A private preference η over resource allocations in X^r is similar to those of a set of traders $K \subset \{1, \dots, H\}$ in the market economy E if:

$$\exists i \in K \text{ and } \exists j \in \{1, \dots, r\} \text{ s.t. } \eta \text{ is similar to } \rho_i \text{ in position } j, \ \forall \xi \in X^r \tag{7}$$

When the preference η is public, we have:

$$\forall j \in \{1, \dots, r\}, \exists i \in K \text{ s.t. } \eta \text{ is similar to } \rho_i \text{ in position } j, \ \forall \xi \in X^r \tag{8}$$

The next step is to define *spaces of preferences P_E* which consist of preferences similar to those of a subset of traders in the market E. The intuitive notion is that of a class of preferences P_E defined over allocations in X^H where each preference in P_E is similar to the preference of some trader i within a subset of traders of E. Formally, consider the space of allocations for $r \geq 2$ individuals, X^r, where $X = R^N$ or $X = R^N_+$. Let j denote possible positions, $j = 1, \dots, r$:

Definition 32. A space of preferences P_E over allocations X^r is said to contain preferences similar to those preferences of a set of traders $K \subset \{1, \dots, H\}$ in the market economy $E = \{X, \rho_i, \Omega_i, i = 1, \dots, H\}$, for $X = R^N$ when

$$\eta \in P_E \quad \Leftrightarrow \quad \eta \text{ is a preference similar to those of the set of traders } K. \tag{9}$$

The space of preferences P_E consists therefore of either private or public preferences over allocations in X^r; in either case, its preferences are similar to those of the subset K of traders in the market economy E. Note that the preferences in a space P_E need not be increasing even when all the preferences in the market are increasing. Similarity only requires that the gradients of preferences in P_E be in the market cones of some preference in the set of traders K, and the market cones of increasing preferences may contain vectors which are not positive.

The following preparatory lemma describes the geometrical structure of spaces of preferences P_E which are similar to those in a set K of traders of the economy E.

[25] Respect of unanimity then means that if all individuals have the same gradient over all possible private choices, then the social preference's gradient over allocations has the same gradient as the traders in each position, for all possible allocations.

Lemma 33. *Consider a market economy $E = \{X, \rho_i, \Omega_i, i = 1, \dots, H\}$, where $X = R^N$. Let P_E be a space of preferences over allocations for r individuals X^r which are similar to those of a set $K \subset \{1, \dots, H\}$ of traders in E. Then at each allocation $\xi \in X^r$ and for each position $j = 1, \dots, r$, the normals to the indifference surfaces of all preferences in P_E define the set*

$$N^K(P_E) = \left\{ \bigcup_{i \in K} D(\rho_i, \Omega_i) \right\} \cup \{0\} = N_K \cup \{0\} \subset R^N. \tag{10}$$

Furthermore, when every two-trader sub economy of E satisfies limited arbitrage, then $\forall K \subset \{1, \dots, H\}$ the set N_K is connected. The set N_K is contractible $\forall K \subset \{1, \dots, H\}$ if and only if the market economy E satisfies limited arbitrage.

Proof. Since P_E consists of preferences over allocations in X^r which are similar to those of the subset K of traders in the economy E, and these preferences may be private or public, then by definition (9), (7), (8) and (6), at each $\xi \in X^r$ the gradients in the set $N^K(P_E)$ define the set

$$\left\{ \bigcup_{i \in K} D(\rho_i, \Omega_i) \right\} \bigcup \{0\} \subset R^N.$$

The next step is to establish that the set N_K is connected. The condition of *limited arbitrage* for any subset of two traders in the statement of this Lemma, is that is that for any two traders l, j in E

$$D(\rho_l, \Omega_l) \bigcap D(\rho_j, \Omega_j) \neq \emptyset.$$

This implies that the set

$$N_K = \bigcup_{i=1}^{r} D_i(\rho_i, \Omega_i)$$

is connected.

Finally we study the contractibility of the set N_K. Theorem 36 in the Appendix establishes that if $\{C_j\}_{j=1,\dots,J}$ is a family of convex sets and $L \subset J$ then

$$\bigcup_{j \in L} C_j \text{ is contractible } \forall L \subset J$$
$$\Leftrightarrow \quad \forall L \subset J, \bigcup_{j \in L} C_j \neq \emptyset. \tag{11}$$

Since the market cone $\forall_j \, D(\rho_j, \Omega_j)$ is a convex set, the second statement in (12) is equivalent to the condition that E has limited arbitrage. The first statement, in turn, states that the set N_K is contractible. Therefore Theorem 36 in the Appendix implies that limited arbitrage is satisfied if and only if $\forall K \subset \{1, \dots, H\}$, the set N_K is contractible. ■

The following result links the resolution of the social allocation problem with the resolution of the market allocation problem. A minimal restriction, that every two traders have limited arbitrage, is imposed from now on to eliminate somewhat pathological economies where no two traders can reach a competitive equilibrium, or those economies where the spaces of preferences similar to those of the traders are discrete or disconnected:

Assumption (C1): *every two-trader sub economy of E has limited arbitrage.*

Note that condition (C1) is not necessary for the equivalence of limited arbitrage and the existence of a competitive equilibrium. The role of this condition is to ensure that the set of gradients N_K is connected: this was shown in Lemma 33. (C1) is rather mild: it certainly does not imply that N_K is contractible, nor that the economy E satisfies limited arbitrage, nor that E has a competitive equilibrium. For example, Figure 5 above illustrates an economy where every two market cones intersect and thus (C1) is satisfied, but the economy of Fig. 5 does not satisfy limited arbitrage, and it does not have a competitive equilibrium.

It seems useful to explain intuitively the role of *limited arbitrage* in the existence of social choice rules. A condition which is necessary and sufficient for the existence of social choice rules is that the space of preferences should be contractible, Chichilnisky and Heal [33]. Therefore to show the existence of social choice rules on preferences similar to those of the market E we must show that the space of such preferences is contractible. This we do using the condition of limited arbitrage and Theorem 36 in the Appendix.

Limited arbitrage is the non-empty intersection of market cones. However, Theorem 36 proves that the dual cones intersect if and only if their union is contractible. And, by Lemma 34, the union of the market cones is precisely the space where the gradients of the preferences similar to those of the market "live". In other words: the proof that limited arbitrage is necessary for the existence of a social choice rule derives from the results of Lemma 34 above, from Theorem 1 of Chichilnisky and Heal [33], and from Theorem 36 in the Appendix. It may be worth mentioning that preferences which are similar to those in the market E are not necessarily increasing. Furthermore, a space of preferences consisting exclusively of increasing preferences may not admit social choice rules because the union of the market cones may have "holes" and therefore may fail to be contractible. Figure 5 illustrates this.

8.1.1. Contractibility and similarity

Consider a space P_E of preferences which are similar to those of the traders in E. From Chichilnisky and Heal [33] we know that a social choice rule on this space of preferences exists if and only if the space of preferences P_E is *contractible*, as defined in Section 2. This condition of contractibility simply means that there exists a continuous way of deforming the preferences through the space P_E, so that at the end of this process we have complete unanimity. With contractibility of the space χ we are assured of the existence of a social choice map, and therefore we are assured of a resolution to our resource allocation problem.

But contractibility is a restriction on social diversity, no more and no less. It tests whether there is a way of deforming continuously our space of individual preferences into itself so that at the end of this deformation all the individuals have identical preferences. A discussion of the role of contractibility in limiting social diversity and in public decision making is in Heal [44]. Thus we are back at the source of the problem of resource allocation in markets: individual diversity.

9. Social allocations in a market

We now specialize the social choice problem to a market economy $E = \{X, \rho_i, \Omega_i, i = 1, \dots, H\}$. Let P_E be a space of preferences over resource allocations in X^r, $r \geq 2$, consisting of preferences which are similar to those of K traders in the economy E as defined in (9). A social choice map ϕ assigns to each profile of r individual preferences in X^r another preference in P_E, the social preference over allocations, $\phi(\kappa_1, \dots, \kappa_r) = \theta$. Each κ_j is a preference similar to the preference of an individual i in E and so is $\theta = \phi(\kappa_1, \dots, \kappa_r)$.

9.1. Example: a classical social welfare function

Each individual $i = 1, \dots, H$ has a utility function $u_i : X \to R$, where $X = R^N$. The social preference W over allocations in X^H is defined for each allocation $(x_1, \dots, x_H) \in X^H$ by

$$W(x_1, \dots, x_H) = \sum_{i=1}^{H} u_i(x_i). \tag{12}$$

The gradient of the function $W : R^{N \times H} \to R$ at an allocation $(x_1, \dots, x_H)$ is a vector in $R^{N \times H}$, while each individual utility u_i has a gradient in R^H at the vector x_i. In particular, the gradient of the function W is *not* the sum or any convex combination of the gradients of the individuals $i = 1, \dots, H$.

The construction of the social welfare function W of (12) is cardinal, in the sense that it is defined over profiles of individual *utilities*, and yields a social *utility* function $W(x_1, \dots, x_H)$. However, the social welfare function of (12) can be used to define a social choice map in ordinal terms, namely a map from profiles of individual preferences to social preferences. Consider a profile of private preferences $(\kappa_1, \dots, \kappa_H)$, where κ_i is a private preference over allocations in X^H. Then $\forall i\ \kappa_i$ is induced by a utility function over private consumption $u_i(\kappa_i) : X \to R$. We may now define the social preference over allocations in X^H corresponding to the profile of preferences $(\kappa_1, \dots, \kappa_H)$, the preference $\Theta(\kappa_1, \dots, \kappa_H)$, as follows. The normal of $\Theta(\kappa_1, \dots, \kappa_H)$ at an allocation $\xi = (x_1, \dots, x_H) \in X^H$ is defined as the vector

$$N\Theta(\kappa_1, \dots, \kappa_H)(\xi) =$$

$$(\lambda_1 G_1 W(u_1(\kappa_i), \dots, u_H(\kappa_H))(\xi), \dots, \lambda_H G_H W(u_1(\kappa_i), \dots, u_H(\kappa_H))(\xi)), \tag{13}$$

where $G_j W$ is the normalized gradient of $W(x_1, \dots, x_H)$ with respect to x_j, i.e.

$$G_j W(\xi) = (\partial W / \partial x_j)(\xi) \in R^N,$$

and $\forall j$, $\lambda_j \in R_+$. The role of the real numbers λ_j in (13) is to normalize the right hand side of the expression (13) so that it defines a vector of length one as it corresponds to the normal of a smooth preference defined in Section 2 [36]. Expressions (12) and (13) define a map from profiles of private preferences over allocations in X^H, into preferences over allocations in X^H:

$$(\kappa_1, \ldots, \kappa_H) \to \Theta(\kappa_1, \ldots, \kappa_H). \tag{14}$$

Clearly, the definition in (14) is the ordinal version of the cardinal construction in (12).

The interest of the social welfare function Θ defined in (14) is that it provides an example of the type of structure for social choice with which we work in this paper. The similarities are: Θ is defined over profiles of *private* preferences over allocations, and it assigns to such a profile another preference over allocations which is typically a *public* preference. If the preferences in the profile $(\kappa_1, \ldots, \kappa_H)$ are those of a market economy $E = \{X, \kappa_i, \Omega_h, i = 1, \ldots, H\}$, then the social preference assigned to profile $\Theta(\kappa_1, \ldots, \kappa_H)$ is similar to those of the traders in E as defined in (9) above. Furthermore, the map Θ is continuous on its arguments and it respects unanimity. Therefore the map Θ satisfies many of the properties required of social choice functions.

The differences with our framework are as follows: firstly, Θ is defined solely on private preferences, while we allow social choice maps which are more general, defined over both private and public preferences over allocations. Secondly, the map Φ, by its construction, is not anonymous. Indeed, the map Φ assigns individual i a dictatorial power over the ith position in the allocation.

9.2. Example: the convex addition of gradients

As pointed out in 9.1, the gradient of the classical welfare function W defined in (12), (13) is neither the sum nor the convex combination of the individual utilities' gradients. A natural question is whether a construction based on the sum or a convex combination of the individual utilities's gradients could yield an adequate social preference. Such construction would have the advantage that it is defined generally, without reference to additional conditions such as limited arbitrage. The following example shows that such a construction, although appearing to be natural, typically does not work. Indeed we argue below that it only works properly when the limited arbitrage condition is satisfied.

Consider for example an economy with two agents each with a linear utility defined on the consumption set $X = R^2$. The two utilities are different. The global cones A_1, A_2 of the two linear preferences are two different half spaces, and the corresponding market cones D_1, D_2 are the two gradient vectors indicated in Fig. 9.

Since the two vectors are different, the market cones do not intersect, and limited arbitrage fails. Now consider the linear preference ρ over allocations in $R^2 \times R^2$ with gradient (D_3, D_3), where D_3 is a convex combination of D_1 and D_2 as shown in Fig. 10, $D_3 = \lambda D_1 + (1 - \lambda) D_2$, $0 < \lambda < 1$.

A problem arises in that although the initial endowment allocation (Ω_1, Ω_2) is strictly preferred by both individuals to the allocation denoted (θ_1, θ_2) in Fig. 10, the social preference ρ strictly prefers (θ_1, θ_2) to (Ω_1, Ω_2) instead.

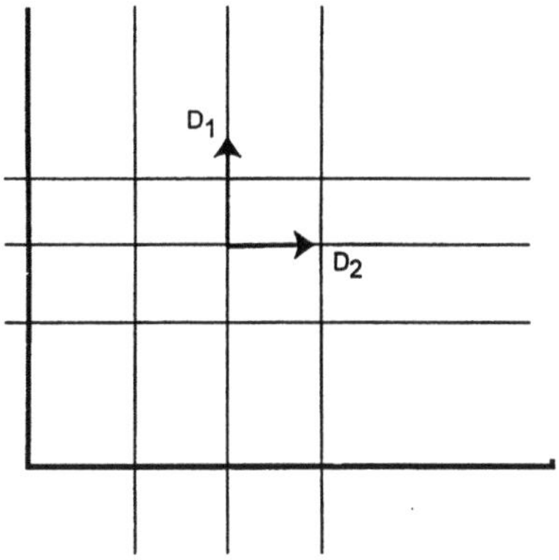

Fig. 9. Two linear preferences with market cones D_1 and D_2

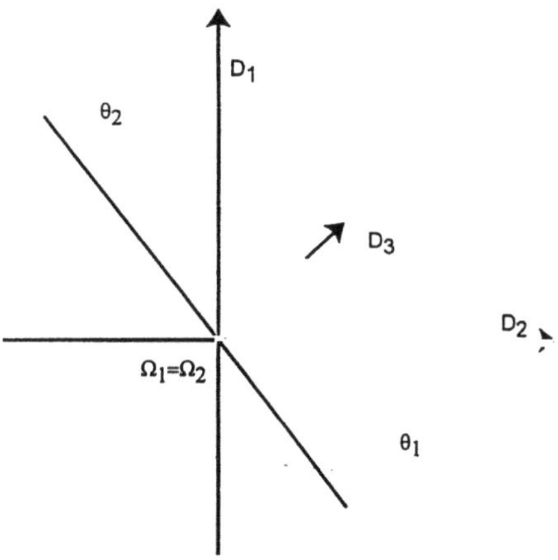

Fig. 10. A linear preference with market cone D_3

This property of ρ contradicts the standard Pareto condition which requires that when everyone strictly prefers a given allocation to another, so should the social preference. The reason for this contradiction is that the two dual cones do not intersect. When the dual cones intersect, then a vector in this intersection defines the gradient of a linear preference which respects the Pareto condition: this follows directly from the definition of dual cones. Otherwise, no preference exists respecting the Pareto condition.

Another way of looking at the same problem is that the social preference defined by the addition of the two gradients in not *similar* to the preferences with gradients D_1 and D_2, as defined in Section 8. Therefore the addition of gradients cannot define a map from profiles of similar preferences into similar preferences as required. This example shows that the convex sum of gradients cannot be used generally as a way of defining appropriate social choice rules: the preference with gradients $D_3 = \lambda D_1 + (1 - \lambda) D_2$, $0 < \lambda < 1$, is not an adequate social preference for the profile (D_1, D_2).

The only possibility of respecting the similarity of preferences or the Pareto condition, would be to assign the social preference with gradient equal to either D_1 or to D_2 to the profile (D_1, D_2). This would respect the similarly condition, and would define a continuous, anonymous social choice map respecting unanimity on the space of preferences $P_E = \{D_1, D_2\}$. But this is possible only because the space P_E is a discrete case consisting only of two elements, the two linear preferences. In the general case in which the space P_E is connected, a continuous anonymous social choice rule satisfying unanimity exists if and only if P_E is contractible, as proved in Chichilnisky and Heal [33]. And, as shown in the Appendix Theorem 9 and Corollary 2, limited arbitrage is actually equivalent to the contractibility of the spaces of preferences.

Our next step is to extend and refine the classical welfare function W defined in (12). The welfare function W is a cardinal construct, while we wish to provide an ordinal one: a social choice map defined over profiles of private and public preferences over allocations, preferences which are similar to those of a market economy E. We seek an anonymous social choice map, a requirement that neither W in (12) nor its ordinal counterpart Θ in (14), satisfy. Theorem 34 in Section 11 establishes that this task can be accomplished when the preferences in the market economy E satisfy limited arbitrage, and therefore by Theorem 36 and Corollary 37 of the Appendix, when the market E has a competitive equilibrium, and only then.

Assume now that a social choice map ϕ with the desired properties $(A1)(A2)(A3)$ exists. The existence of a social choice map ϕ solves the resource allocation problem from the point of view of social choice:

We say that *the allocation problem is resolved for preferences similar to those of the market* $E = \{X, \rho_i, \Omega_i, i = 1, \dots, H\}$ when for every $r \geq 2$ and any space of preferences P_E over allocations in X^r similar to a set of traders $K \subset \{1, \dots, H\}$ in E there exists a continuous anonymous social choice map $\phi: (P_E)^r \to P_E$ respecting unanimity.

Consider a market economy $E = \{X, \rho_i, \Omega_i, i = 1, \dots, H\}$ having a social choice map $\phi: (P_E)^H \to P_E$ satisfying the axioms $(A1)(A2)(A3)$ on the space of preferences P_E similar to those of the economy E.

A social allocation for E is a feasible allocation of E which is optimal in the space of feasible allocations $\Upsilon = \{x_1, \dots, x_H: \sum x_i = \Omega\}$ according to the social preference $\phi(\rho_1, \dots, \rho_H)$.

10. Comparing the two resource allocations problems

10.1. Existence of market allocations

As seen in Section 3 above, even when individual preferences are smooth, concave and increasing, and when the consumption sets are positive orthants, Arrow-Debreu economies may have no competitive equilibrium. The problem is quite general and it occurs in economies with any number of individuals and of goods.[26] Limited arbitrage limits precisely the degree of diversity among the agents of the economy to one at which market equilibrium will exist. Indeed, for economies with consumption bounded below, $X = R^N_+$, such limits on diversity are implicit in Arrow's *resource relatedness* [3] and in McKenzie's *irreducibility* condition [48, 49, 50]. All these conditions ensure that the endowments of any household are desired, directly or indirectly, by others, so that their incomes cannot fall to zero. In this case our limited arbitrage condition is always satisfied. Yet zero income by itself does not rule out the existence of a competitive equilibrium; there may be economies with a competitive equilibrium in which some consumers have zero income. Figure 7 above illustrates. Furthermore an allocation where some individuals have zero income reflects a real problem: the fact that some individuals are considered worthless, that they have nothing to offer that others want. Our condition of *limited arbitrage* brings out the issue of diversity by focusing on the problem of zero income individuals. It does not attempt to rule out the case of individuals with zero income; instead, it seeks to determine if society's evaluation of their worthlessness is widely shared.

10.2. Existence of social allocations

Within the same type of market economy studied above $E = \{X, \Omega_i, \rho_i, i = 1, \dots, H\}$, the *social allocation problem* can be defined as follows. Find a social choice rule ϕ satisfying the required axioms (A1)–(A3): ϕ must be defined on a space of preferences which contains those of the market E. The problem has generally no solution because a social choice rule satisfying the required axioms may not exist, Chichilnisky [12]. Therefore a social contract promising a solution which satisfies the ethical principles agreed and having a completely universal domain of preferences, may not deliver. Additional conditions or preferences are needed. Here, as in the case of the market equilibrium, the conditions should be on the exogenous data which identify the economy E, namely on the individuals' endowments Ω_i and their preferences ρ_i.

---

[26] Other concepts of market equilibrium could be utilized to define a market allocation, such as for example the *quasi-equilibrium* or *pseudoequilibrium* introduced by Debreu [38], or as defined in Arrow and Hahn [3]. These are closely related definitions; they define allocations where individuals minimize cost rather than maximizing utility. When individuals' incomes are strictly positive, these concepts agree with the competitive market equilibrium (Arrow and Hahn [3], Chapter 4) but as Arrow and Hahn point out, the condition that all individuals should have strictly positive endowments of all goods is unrealistic ([3], Chapter 4, p. 80). *Quasi-equilibrium* allocations have the advantage that they always exist with continuous concave preferences and with positive orthants as consumption sets. However, the competitive equilibrium stands alone in terms of its welfare properties; quasi-equilibrium allocations are not generally Pareto efficient. Therefore the main justification for using market allocations, which is efficiency, is lost. For this reason we concentrate here on competitive equilibrium allocations. For a study of limited arbitrage and quasiequilibrium, see Chichilnisky [26].

We know that the space of all smooth preferences over allocations in X^H is too large: the three axioms of continuity, anonymity and respect of unanimity are inconsistent in that case and no social choice rule satisfying them exists, Chichilnisky [12]. We must therefore search for a universe of preferences and endowments in which a social choice rule ϕ does exist. Such a universe must contain the preferences of our economy E, and we shall require that it should consist of individuals who share the essential characteristics of the economy E. The motivation is to compare the types of restrictions needed for existence of a competitive equilibrium with those required for the existence of a social choice map.

11. Linking markets, arbitrage and social choice

This section provides the main results linking resource allocation by markets and by social choices.

Theorem 34. *Consider an economy* $E = \{X, \rho_i, \Omega_i, i = 1, \dots, H\}, H \geq 2, X = R^N$ $N \geq 1$ *satisfying* (C1).

The following properties are equivalent

(*a*) *The market economy E has a competitive equilibrium*

(*b*) *Every sub economy of E with at most* $N + 1$ *traders has a competitive equilibrium*

(*c*) *E has limited arbitrage*

(*d*) *E has limited arbitrage for any subset of individuals with no more than* $N + 1$ *members.*

(*e*) *For any space of preferences* P_E *similar to those of a subset K of market traders, there exists a continuous anonymous social choice map* $\phi:(P_E)^r \to P_E$ *respecting unanimity, for all* $r \geq 2$, *and*

(*f*) *The economy E has a nonempty core.*

Proof. The proof that (a)$\Leftrightarrow$(c) is in [18, 20, 23, 24, 26]. Next I establish the equivalence between (c) and (e).

Case 1. The consumption set $X = R^N$. We first show that limited arbitrage is a necessary condition for the existence of the social choice map ϕ for all $K \subset \{1, \dots, H\}$ and all $k \geq 2$ satisfying the three axioms (A1)–(A3).

Assume that such a social choice map ϕ exists for all $r \geq 2$. Let LP_E be the subspace of P_E consisting of all its linear preferences, i.e. those preferences within the space P_E which are representable by linear utility functions having their gradients, in each position $j = 1, \dots, r$, contained in the set $N_K \cup \{0\} \subset R^N$. Let $in: LP_E \to P_E$ denote the inclusion map. Note that each preference $v \in LP_E$ is uniquely identified by the normal $Nv \in R^{Nr}$ to one of its indifference surfaces: by linearity indifference surfaces have the same normals at every allocation. Therefore

$$LP_E \approx (N_K \cup \{0\})^r, \tag{15}$$

i.e. the space of linear preferences LP_E is homeomorphic to the product space $(N_K \cup \{0\})^r$. Now consider an allocation $\xi \in X^K$ and define the map

$\pi : P_E \to LP_E$ so that

$$\pi(\kappa) = \kappa\xi,$$

where $\kappa\xi \in LP_E$ is the linear preference over allocations in X^r having as its gradient in R^{Nr} the vector $G\kappa(\xi)$ which is the normal to the indifference surface of the preference κ at ξ. Both maps *in* and π are continuous on their domains.

By assumption, there exists a social choice map $\phi:(P_E)^r \to P_E$ for any $r \geq 2$, satisfying the three axioms. Now consider for any $r \geq 2$ the map induced by the composition of *in* and ϕ, defined by

$$\psi:(LP_E)^r \to LP_E,$$

$$\psi(v_1, \ldots, v_r) = \pi(\phi(in[v_1], \ldots, in[v_r])).$$

The map $\psi:(LP_E)^r \to LP_E$, is continuous, anonymous and respects unanimity by construction, because ϕ satisfies these three properties. Since by (15) $\forall j$, $LP_E \approx (N_K \cup \{0\})^r$, ψ defines a map $\Lambda:[(N_K \cup \{0\})^r]^r \to (N_K \cup \{0\})^r$ for all $r \geq 2$ satisfying the three axioms (A1)–(A3). Since Λ is continuous and respects unanimity, it maps each connected component of the space $(N_K \cup \{0\})^r$ into itself. In particular, the restriction of the map Λ to the connected component $[(N_K)^r]^r$ of $[(N_K \cup \{0\})^r]^r$, denoted $\Lambda/(N_K)^r$, maps $[(N_K)^r]^r$ into $(N_K)^r$, i.e. $\Lambda/(N_K)^r:[(N_K)^r]^r \to (N_K)^r$, and it satisfies the three axioms (A1)(A2)(A3). By Theorem 1 of Chichilnisky and Heal [33] such a map Λ exists if and only if the space $(N_K)^r$ is contractible for all $r \geq 2$; this in turn is true if and only if the space N_K is contractible. Therefore the contractibility of the space N_K is necessary for the existence of the social choice map ϕ. But Theorem 37 in the Appendix proves that N_K is contractible for all $K \subset \{1, \ldots, H\}$ if and only if the *limited arbitrage* condition is satisfied. This completes the proof of necessity of limited arbitrage.

We now turn to the proof of sufficiency. By definition, limited arbitrage implies the existence of a non-zero vector $v \in R^N$ in the intersection of all the market cones:

$$v \in \bigcap_{i=1}^{H} D_i(\rho_i, \Omega_i).$$

Furthermore, by definition of the space P_E, all the indifference surfaces of any preference in P_E must intersect the ray defined by the vector $(v, \ldots, v)$ in the space of allocations X^{Nr}. Therefore the conditions required in Chichilnisky [11] the existence of a social choice map $\phi:(P_E)^r \to P_E$ satisfying the three axioms (A1)–(A3) are satisfied for the space P_E. This completes the proof of sufficiency of limited arbitrage. Therefore (c)$\Leftrightarrow$(e).

The proofs that (a)$\Leftrightarrow$(b) and (c)$\Leftrightarrow$(d) follow directly from Theorem 37 in the Appendix. The proof that (c)$\Leftrightarrow$(f) follows from [20, 21, 23, 28]. ■

Theorem 36. *Consider a market economy $E = \{X, \rho_i, \Omega_i, i = 1, \ldots, H\}$. Any market allocation $(x_1^*, \ldots, x_H^*)$ of E is also a social allocation for E.*

Proof. When a market allocation for E exists, E has a competitive equilibrium described by a price vector p^* and an allocation $(x_1^*, \ldots, x_H^*) \in X^H$ which is individually optimal within the budget sets and which clears all the markets. We shall show that this market allocation $(x_1^*, \ldots, x_H^*)$ is also *social allocation.*

This means that any space of preferences P_E defined over allocations for the H traders, namely on X^H, and consisting of preferences which are similar to those of the H traders in the economy $E = \{X, \rho_i, \Omega_i, i = 1, \ldots, H\}$, admits a social choice map $\Psi:(P_E)^H \to P_E$ satisfying the axioms (A1)(A2)(A3), and such that $(x_1^*, \ldots, x_H^*)$ optimizes the social preference $\Psi(\rho_1, \ldots, \rho_H)$ over the set of all feasible allocations Υ of the economy E.

Since a market equilibrium exists, by Theorem 34 there exists a social choice map $\phi:(P_E)^H \to P_E$ satisfying the three axioms (A1)–(A3), for the space of preferences P_E similar to the H traders in the economy E. By Theorem 34 we also know that limited arbitrage is satisfied, i.e. $\bigcap_{i=1}^H D(\rho_i, \Omega_i) \neq \emptyset$.

We now use a partition of unity on the space P_E as defined in Appendix 0 in order to define a modification of the social choice map ϕ, called Ψ, which also satisfies the three axioms, and according to which the equilibrium allocation is optimal within all feasible allocations for the social preference $\Psi(\rho_1, \ldots, \rho_H)$. By the results of Chichilnisky [11] the space P_E is contained in a manifold which is the inverse image under a smooth retraction of a linear space, the space spanned by the vector $(v, \ldots, v) \in R^{N \times H}$, where $v \in \bigcap_{i=1}^H D(\rho_i, \Omega_i)$. Therefore the space P_E is Hausdorff and we can therefore apply a partition of unity, see the Appendix.

Let U be the set of preference profiles in P_E consisting of the profile $(\rho_1, \ldots, \rho_H)$ and of all its permutations,

$$U = \bigcup \pi\{(\rho_{\pi_1}, \ldots, \rho_{\pi_H})\}, \text{ for all permutations } \pi \text{ of the set } \{1, \ldots, H\}.$$

The set U consists of finitely many points in $(P_E)^H$; U is disjoint from the diagonal $\Delta(P^E)^H = \{(\kappa_1, \ldots, \kappa_H) \in (P^E)^H : \forall i, j\ \kappa_i = \kappa_j\}$, because the profiles in U consist of private preferences each defined over a different position. Since $(P_E)^H$ is contained in a manifold, it is a Hausdorff space.

We now construct a new social choice map Ψ with the desired properties. Using a partition of unity we modify the map $\phi:(P_E)^H \to P_E$ to obtain another continuous anonymous map $\psi:(P_E)^H \to P_E$, which differs from ϕ only in an open neighborhood $\theta(U)$ of the set U, and is otherwise identical to ϕ. Within the set U the new map Ψ satisfies:

$$\Psi(\rho_1, \ldots, \rho_H) = \Psi(\rho_{\pi_1}, \ldots, \rho_{\pi_H}) = \kappa_{p^*} \tag{16}$$

for all permutations π of the set of indices $\{1, \ldots, H\}$, where in (16) κ_{p^*} is the linear preference over allocations in X^H with gradient vector $(\lambda_1 Du_1(x_1^*), \ldots, \lambda_H Du_H(x_H^*))$, for some vector $(\lambda_1, \ldots, \lambda_H) \in R_+^H$; $Du_i(x_i^*)$ is the gradient of u_i at (x_i^*) for a utility u_i which represents ρ_i, and

$$\Psi(\kappa_1, \ldots, \kappa_H) = \phi(\kappa_1, \ldots, \kappa_H).$$

if $(\kappa_1, \ldots, \kappa_H) \notin \theta(U)$. The map

$$\Psi:(P_E)^H \to P_E$$

satisfies the three axioms by construction. We now show that the competitive equilibrium allocation $(x_1^*, \ldots, x_H^*) \in X^H$ is a social allocation for the economy E with the social choice map Ψ. Since the allocation $(x_1^*, \ldots, x_H^*)$ is by assumption the equilibrium allocation corresponding to the price vector p^*, there exists a vector $(\lambda_1^*, \ldots, \lambda_H^*) \in R_+^H$, such that

$$W(x_1^*, \ldots, x_H^*) = \mathrm{Max}_{(x_1, \ldots, x_H) \in \Upsilon} W(x_1, \ldots, x_H) \tag{17}$$

where

$$W(x_1, \ldots, x_H) = \sum_{i=1}^{H} \lambda_i^* u_i(x_i),$$

and where the utility $u_i: X \to R$ represents the preference ρ_i. Now choose the vector $(\lambda_1, \ldots, \lambda_H)$ in the definition of Ψ following (16) to be the vector $(\lambda_1^*, \ldots, \lambda_H^*)$ in (17). Then, with this definition of $(\lambda_1, \ldots, \lambda_H)$, the allocation $(x_1^*, \ldots, x_H^*)$ maximizes the social preference $\Psi(\rho_1, \ldots, \rho_H)$ over the feasible set Υ. The allocation $(x_1^*, \ldots, x_H^*) \in X^H$ is therefore a social allocation for the social choice map Ψ and the economy E, as we wished to prove. ■

12. Social diversity and a topological invariant for markets

As defined above, the economy is **socially diverse** when $\bigcap_{h=1}^{H} D_h = \phi$, see also Chichilnisky [26]. Diversity admits different "shades", which can be measured by the smallest number of market cones which do not intersect. I showed in [26] that the resource allocation properties of the economy E can be described simply in terms of a topological invariant, which is given by a family of cohomology rings denoted CH(E).

13. Related literature

13.1. Comparison with Arrow's axioms

It seems worth comparing our axioms with Arrow's [2] axioms of social choice. Arrow's axioms are more suitable for finite set of choices, such as voting among a finite set of n candidates. Instead, we are choosing here among an infinite set of choices, namely among the set Υ of all feasible allocations in Euclidean space.

The three axioms (A1)–(A3) are different, indeed not comparable, to Arrow's [2]. The *anonymity* condition (A2) is stronger than Arrow's *non-dictatorship* axiom, because the former requires equal treatment while the latter eliminates only extreme inequality of treatment. *Respect of unanimity* (A3) is strictly weaker than his *Pareto condition*, since respect of unanimity is only binding when all preferences within a profile are identical. Instead, the Pareto condition applies to any profile of preferences which, equal or not, prefer a given choice x to another y. Finally, Arrow does not consider continuity (A1) as we do, and we do not require Arrow's axiom of *independence from irrelevant alternatives*, an axiom which has been somewhat controversial. In other words: neither set of axioms implies the other.

Continuity is required here on the grounds of statistical tractability: it implies that the sampling of populations' preferences will approach the true distribution provided the grid of observations in the sample is fine enough. The continuity axiom makes this formulation of social choice better suited to continuous sets of choices and to connected sets of preferences. This is because when the space of preferences is discrete or finite – as it would be when there are finitely many choices – then the space of preferences is itself finite and therefore continuity is a vacuous requirement. Recent work by Baigent [5, 6] and by Nitzan [53] has extended the axiom (A1) of continuity to one of "proximity" of preferences, a concept which is appropriate for discrete spaces of preferences and for preferences over finitely many choices. Using "proximity" instead of "continuity", and preserving the other two axioms – anonymity and respect of unanimity – Baigent [5, 6] proved the impossibility results of Chichilnisky [12] for the case of finitely many choices. Chichilnisky [14] has recently shown the connection between the existence of social choice rules and the manipulation of non cooperative games.

What makes the axioms used in this paper particularly well suited for our problem is that they lend themselves naturally to the study of preferences and choices similar to those which are studied in market economics. Furthermore, with these axioms there exist simple necessary and sufficient conditions for resolving the social choice paradox, Chichilnisky and Heal [33]; this is not true for Arrow's axioms.

Recent results by Baryshnikov [7] highlight a topological connexion between Arrow's impossibility theorem and my own in [12, 13]. This is a somewhat surprising result: using algebraic topology, [7] shows that the condition of independence of irrelevant alternatives can be used to prove that the 'nerve' of a family of sets naturally associated with all preferences over a finite set of choices, has the topological structure of a sphere. Therefore my non-existence theorem [12, 13], which applies in particular to spheres, implies Arrow's impossibility theorem in this context.

14. Conclusions

On the basis of previous results [18, 20, 22], and [23], I established that limited arbitrage is a necessary and sufficient condition for the existence of social choice maps on spaces of preferences similar to those of a market economy. The same condition – *limited arbitrage* – is necessary and sufficient for the existence of a competitive equilibrium and the core in a market economy [16, 18, 21, 22, 23, 24, 26, 28]. Furthermore, I showed that a market allocation is always a social allocation. In this sense, the results of this paper unify three forms of resource allocation – by markets, games and by social choice – which have developed and remained separate until now.

I have chosen competitive equilibrium – rather than other concepts of market equilibrium – because of its Pareto efficiency. This property is generally lost in weaker forms of market equilibrium, such as quasi-equilibrium and compensated equilibrium.

I have concentrated on three axioms of social choice: continuity, anonymity and respect of unanimity, introduced in [12], which are particularly well suited for problems where the choices are elements in Euclidean space, and for which necessary and sufficient conditions exist for resolving the social choice problem [33]. Arrow's [2] classic axioms appear to be better suited for problems with finitely many choices. As already pointed out, however, it is possible to draw a close topological link between Arrow's theorem and the impossibility theorem in [12], see [7]. I showed that the condition of limited arbitrage characterizes those economies where the traders do not have Condorcet triples (i.e. cyclical behavior), and those where Arrow's impossibility theorem can be resolved, on choices involving large utility values.

The crucial condition in all this is the contractibility of the space of preferences, which is a measure of social diversity [44]. Limited arbitrage is closely connected with contractibility. A key element is the necessary and sufficient conditions for the non-empty intersection of a family of sets in Theorem 5, Chichilnisky [15, 19]. *Limited arbitrage* is defined as the non-empty intersection of the market cones of the traders, which [19] proved is equivalent to a contractibility condition. Thus I have linked *limited arbitrage*, which is needed for the existence of a market equilibrium and the core, with the *contractibility of the space of preferences*, which is needed for a resolution of the social choice problem. The same restriction on social diversity is therefore necessary and sufficient for the successful resolution of all three problems of resource allocation.

Appendix

0. Partitions of unity

Assume that Y is a Hausdorff space [47]. If U, V are two open neighborhoods of $x \in Y$, $U \subset V, U \neq V$, a continuous map $v: Y \to R$ is called a *partition of unity* for Y, U, V, if $\forall y \in Y - V, v(y) = 0$, and $\forall y \in U, v(y) = 1$, and $\forall y \in Y, 1 \geq v(y) \geq 0$.

Such partitions of unity always exist in Hausdorff spaces, [47]. A partition of unity is used to construct new maps from old ones, having specific values. For example, for any $f: Y \to Z$ any $y \in Y$, $z \in Z$, and any open neighborhood U of $y \in Y$, there exists a continuous map $g: Y \to Z$ such that $g(y) = z$ and $g(x) = f(x)\ \forall x \in X - U$. If the map f is symmetric, g can be constructed to be symmetric also.

1. Intersection and contractibility

Theorem 36. *Consider a family $\{U_i\}_{i=1,\ldots,n}$ of convex sets in R^m for any $n, m \geq 1$. Then*

$$\bigcap_{i=1}^{n} U_i \neq \emptyset \text{ if and only if } \bigcap_{j \in J} U_i \neq \emptyset$$

for any subset of indices $J \subset \{1, \ldots, n\}$ having at most $m + 1$ elements.

For a proof see Chichilnisky [15, 19].

Corollary 37. *A market economy E with n traders and m commodities satisfies limited arbitrage if and only if every sub economy of E with at most $m + 1$ traders satisfies limited arbitrage.*

For a proof see Chichilnisky [19].

References

1. Arrow KJ (1951) An extension of basic theorems of classical welfare economics. In Neyman J (ed) Proceedings of the Second Berkeley Symposium on Mathematical Statistics and Probability. Berkeley, University of California Press, pp 507–532
2. Arrow K (1953) Social Choice and Individual Values. Cowles Foundation Monograph, John Wiley
3. Arrow K, Hahn F (1971) General competitive analysis. North-Holland, Amsterdam, 1986
4. Arrow K, Debreu G (1954) Existence of an equilibrium for a competitive economy. Econometrica 22: 264–90
5. Baigent N (1984) A reformulation of Chichilnisky's theorem. Econ Letters: 23–25
6. Baigent N (1987) Preference proximity and anonymous social choice. Q J Econ 162–194
7. Baryshnikov Y (1993) Unifying impossibility theorems: a topological approach. Adv App Math 14: 404–415
8. Black D (1948) On the rationale of group decision making. J Polit Econ 56: 23–24
9. Black D (1948a) The decisions of a committee using a simple majority. Econometrica 16: 245–61
10. Chichilnisky G Manifolds of preferences and equilibria (1976) PhD dissertation, University of California, Berkeley
11. Chichilnisky G (1986) Topological complexity of manifolds of preferences. In: Hildenbrand W, Mas-Colell A (eds) Essays in Honor of Gerard Debreu. North-Holland, Amsterdam, Chapter 8, 131–142
12. Chichilnisky G (1980) Social choice and the topology of spaces of preferences. Adv Math 37(2): 165–176
13. Chichilnisky G (1982) Social aggregation rules and continuity. Quarterly Journal of Economics 97: 337–352
14. Chichilnisky G (1993) On strategic control. Q J Econ 108: 285–290
15. Chichilnisky G (1981) Intersecting families of sets. In: Essex Working Papers, University of Essex, U.K.
16. Chichilnisky G (1992) Limited arbitrage is necessary and sufficient for the existence of a competitive equilibrium. Working Paper No. 650, Department of Economics, Columbia University
17. Chichilnisky G (1993) Topology and economics: the contribution of Stephen Smale. In: Hirsch M, Marsden J, Shub M (eds). From Topology to Computation, Proceedings of a Conference in Honor of S. Smale. Springer Verlag, Berlin, Heidelberg, New York, pp 147–161
18. Chichilnisky G (1992) Markets, arbitrage and social choice. Paper presented at the conference 'Columbia Celebrates Arrow's Contributions", Columbia University, New York, October 27, 1991. Working Paper No. 586, Columbia University, December 1992. CORE Discussion Paper No. 9342, CORE, Universite Catolique de Louvain, Louvain la Neuve, Belgium, 1993
19. Chichilnisky G (1993) Intersecting families of sets and the topology of cones in economics. Bull Am Math Society Expository and Research Papers, 29(2): 189–207
20. Chichilnisky G (1995) A unified perspective on resource allocation: limited arbitrage is necessary and sufficient for the existence of a competitive equilibrium, the core and social choice. CORE Discussion Paper No. 9527, Universite Catolique de Louvain, Louvain-la-Neuve, Belgium, April 1995
21. Chichilnisky G (1996) Limited arbitrage is necessary and sufficient for the nonemptiness of the core. Econ Letters (in press) (First published as Limited arbitrage is necessary and sufficient for the existence of an equilibrium and the core and it limits voting cycles. Econ Letters, in an issue which was reprinted without the paper December 1994, 46: 321–331

22. Chichilnisky G (1994) Social diversity and gains from trade: a unified perspective on resource allocation. Am Econ Rev, Papers and Proceedings, 84(4): 427–434
23. Chichilnisky G (1995) A unified perspective on resource allocation: limited arbitrage is necessary and sufficient for the existence of a competitive equilibrium, the core and social choice. (Invited presentation at the International Economic Association, May 1994, Hernstein Schloss, Vienna, Austria. Arrow K, Sen A, Suzumura T, eds. Social Choice Reexamined. McMillan (in press)
24. Chichilnisky G (1995) Limited arbitrage is necessary and sufficient for the existence of a competitive equilibrium with or without short sales. Econ Theory 5(1): 79–108
25. Chichilnisky G (1996) Limited arbitrage is necessary and sufficient for the existence of a competitive equilibrium. J Math Econ December 1996
26. Chichilnisky G (1996) A topological invariant for competitive markets. Working Paper, Columbia University. J Math Econ December 1996
27. Chichilnisky G (1996) Limited arbitrage and uniqueness of equilibrium in strictly regular economies. Working Paper No 9596–21, Columbia University
28. Chichilnisky G (1996) Markets and Games: a Simple Equivalence among the Core, Equilibrium and Limited Arbitrage. Metroeconomica Vol 47, No 3, October 1996, pp 266–280
29. Chichilnisky G, Heal GM (1984) Existence of a competitive equilibrium in L_p and Sobolev spaces. IMA Preprint Series No. 79, Institute for Mathematics and its Applications, University of Minnesota, Minneapolis, MN
30. Chichilnisky G, Heal G (1993) Existence of a competitive equilibrium in Sobolev spaces without bounds on short sales. J Econ Theory 59: 364–84
31. Chichilnisky G, Heal GM (1992) Equilibrium and the core with finitely or infinitely many markets: a unified approach. Econ Theory (in press)
32. Chichilnisky G, Heal GM (1992) Arbitrage and equilibrium with infinitely many securities and commodities. First Boston Paper Series 92–29, July 1992, Columbia Business School, Columbia University
33. Chichilnisky G, Heal G (1983) Necessary and sufficient conditions for a resolution of the social choice paradox. J Econ Theory 31(1): 68–87
34. Condorcet Marquis de (1785) Essai sur l'Application de l'Analyse a la Probabilite des Decisions Rendues a la Pluralite des Voix. Paris
35. Debreu G (1959) The Theory of Value, Cowles Foundation Monograph, John Wiley
36. Debreu G (1954) Competitive equilibrium and Pareto optimum. Proc Nat Acad Sciences 40: 588–92
37. Debreu G (1971) Smooth preferences. Econometrica 40: 603–615
38. Debreu G (1962) New concepts and techniques for equilibrium analysis. Inter Econ Rev 3: 257–273
39. Grandmont JM (1982) Temporary equilibrium. Arrow K, Intriligator M (eds) Handb Math Econ New York, North Holland
40. Green J (1973) Temporary equilibrium in a sequential trading model with spot and futures transactions. Econometrica 41(6): 1103–23
41. Hammond P (1983) Overlapping expectations and Hart's conditions for equilibrium in a securities market. J Econ Theory 31: 170–75
42. Negishi T (1960) Welfare economics and the existence of an equilibrium for a competitive economy. Metroeconomica Vol 2, 92–97
43. Hart O (1974) Existence of equilibrium in a securities model. J Econ Theory 9: 293–311
44. Heal GM (1983) Contractibility and public decision making. In: Pattanaik P, Salles M, (eds) Social Choice and Welfare, Chapter 7, North-Holland
45. Helly E (1933) Uber Mengen Konvexen mit Gemeinschlaftlichen Punkten. J Deutsch Math Verein 32: 175–186

46. Kant (1788) Critike der pratischen Vernunft. English translation by L. W. Beck Critique of Practical Reason. Liberal Arts Press, New York, 1956
47. Kelley J (1960) General Topology. Van Nostrand, New York
48. McKenzie L (1959) On the existence of a general equilibrium for competitive markets. Econometrica 27: 54–71
49. McKenzie L (1987, 1989) General equilibrium. Chapter 1, General Equilibrium, The New Palgrave (eds. Eatwell J, Milgate M, Newman P) Norton, New York
50. McKenzie L (1961) On the existence of general equilibrium: some corrections. Eonometrica 29: 247–248
51. Monteiro P, Page F, Wooders M (1996) Arbitrage and gains from trade: a counter-example. J Math Econ (in press), mimeo
52. Nielsen L (1989) Asset market equilibrium with short selling. Rev Econ Studies 56(187): 467–473
53. Nitzan S (1989) More on the preservation of preference proximity and anonymous social choice. Q J Econ: 187–190
54. Page FH Jr. (1987) Notes and comments to the editor: on equilibrium in Hart's securities exchange model. J Econ Theory 41(2): 392–404
55. Pattanaik P, Sen A (1969) Necessary and sufficient conditions for rational choice under majority decisions. J Econ Theory 1: 178–202
56. Spanier E (1966) Algebraic Topology. McGraw Hill
57. Werner J (1987) Arbitrage and the existence of competitive equilibrium. Econometrica 55(6): 1403–1418

Topological and discrete social choice: in a search of a theory

Yuliy M. Baryshnikov

Department of Mathematics, Osnabrück University, 49069 Osnabrück, Germany

Received: 30 October 1994/Accepted: 22 April 1996

Abstract. In this note we discuss a possibility of introducing topological methods in the combinatorial paradigm of the social choice theory.

1. Introduction

This paper demonstrates a remarkable interplay between two theories of social choice – a *topological* one, initiated and developed mainly by Chichilnisky (see, e.g. [3–5]), and the *classical, combinatorial* one, stemming from the work of Arrow [1–11]. Both theories deal with the aggregation of preferences – with apparently cardinally different notions of preferences. Recall that in the classical theory, the preferences are assumed to be given on discrete sets of alternatives and constitute a discrete set by themselves. This bounds the technique of the theory to be combinatorial. In the topological theory of social choice, the set of alternatives is assumed from the beginning to have the structure of a topological space, say a manifold embedded in Euclidean space. The preferences on such spaces are usually (smooth) functions modulo order preserving transformations in the target real line, that is smooth preferences in the sense of Debreu [9]. The totality of these preference relations also can be given the structure of a topological space. This theory is well developed in the context of the differentiable approach to the general equilibrium problem [10].

Until recently it has been implicitly assumed that the theories coexist but do not have much in common. The combinatorial setting was considered as

The author benefited very much from discussions with G. Chichilnisky of results presented here. The comments and support of G. Heal is heartily appreciated. This work was supported by the Deutsche Forschunggemeintschaft under grant Ba86/1–2

primary and more natural and simple, while the topological one was usually seen as hi-tech fortresses with no life nearby.

The aim of this note is to demonstrate that both theories are in fact much closer to each other than was commonly thought. Actually, I believe that they are in fact two different guises of the same theory which uniformly covers both discrete and continuous phenomena of the social choice theory.

At the current stage I do not have this Grand Unifying Theory at hand – and do not think it is reasonable to hurry to formulate it. To have a good general theory one have to have a lot of evidences of its existence, and a lot of working examples and well worked out technical aids.

That is what is done here – I give two examples of *topologizing combinatorial problems of social choice.* In one case, it just gives an easy solution of what otherwise could need a lot of intricate logical reasonings. In other case, we will give a topological proof of the true jewel of the social choice theory – Arrow's impossibility theorem.

Chichilnisky [7, 8] has recently approached the same issue from a very different perspective. She has shown that, for choices involving large utility values, her formulation and Arrow's are equivalent (see also Chichilnisky's paper in this volume and Heal's Introduction).

I will not give any proofs here. For the first model the technicalities are either easy or can be found in the literature; in the second case the proofs are given in my paper [2].

I use here some topological concepts. Actually they are much more elementary than one might think, but they certainly need to be looked up in a primer on algebraic topology. The only concept I would like to spell out here is the idea of *the nerve* because this is somehow always on the fringe of the basic topology courses and thus a reader comfortable with homologies and the likes can suddenly stumble upon it. The concept is very simple and useful however, and will be actively used in what follows (as it is in [6]).

Consider a covering of a set – a number of subjects whose union is the whole set (it does not matter whether or not the set is a topological space). A nerve associated with this covering is a simplicial complex, that is it is a *topological space* made by attaching simplices of different dimensions along their boundary simplices, in the following way: The set of 0-dimensional simplices is just the set of the covering subsets. So each of the covering subsets is a vertex of the simplicial complex. A simplex spanned by a finite number of the vertices belongs to the nerve if and only if the intersection of the subsets is nonempty. So, for example, if two covering sets intersect, then in the simplicial complex the corresponding vertices are joined by a 1-simplex, an edge. And if three covering sets intersect, then the triangle with corresponding points as vertices, that is a 2-simplex, belongs to the simplicial complex, and so on. The Fig. 1 shows some examples.

The nerves were introduced into topology to provide topological spaces with simplicial models thus moving towards a more combinatorial theory (surely, a simplicial complex is much more discrete than a general topological space!) Here we are mostly moving into the inverse direction, using nerves to topologize combinatorial objects.

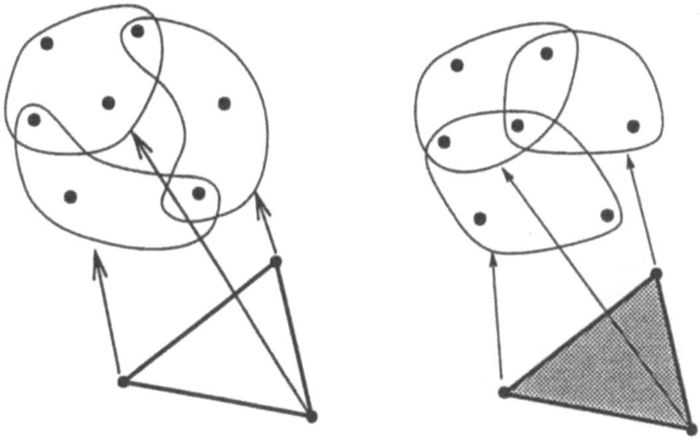

Fig. 1. Examples of nerves of coverings of finite sets

2. A toy model: searching for a camping place and symmetric social rules

Consider the following specific social choice problem. Let X be a finite set of cardinality $n \geq 2$ of camping sites located along a circular road (Fig. 2). A group of hikers is going to choose one of them. Each of the k hikers has his own beloved place to stay, and the social choice problem here is provide them with a procedure for aggregating their preferences.

In other words, the input for the aggregation rule is just a k-tuple of the preferred sites of each voter. The output of the procedure should be a single site, so that an aggregation rule in this case is just a mapping $f\colon X^k \to X$.

The following axioms seem reasonable:

Stability. If a voter suddenly changes his opinion and claims that his favorite site is a site next to his previous choice, then the output of the aggregation rule changes at most to a site next to the previous output. The meaning of this condition is that the aggregated choice is stable under small errors of the voters.

Symmetry. We assume anonymity of the aggregations rule: formally, the rule is invariant under permutations of the incoming data.

Unanimity. If all the voters choose the same pick-nick site, then the aggregation rule also picks this site.

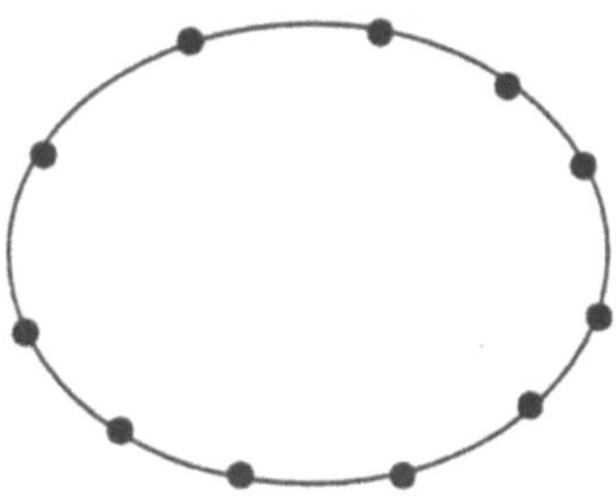

Fig. 2.

The remarkable fact is that if the number of sites is large enough, then there is no rule satisfying all three axioms given above.

Theorem 1. *If $n > 2k$, then there exists no stable symmetric unanimous aggregation rule.*

The proof of the result mimicks the following well-known theorem in the topological social choice theory. Let X be a topological space, and C a class of preferences on X. We assume that the set of such preferences itself is given the structure of topological space (a thorough discussion of possible topologizations in different settings can be found in [3, 9]). We assume further, that the space of preferences has the homotopy type of CW-complex (a common assumption in topological lore to exclude some pathologies; in fact anything reasonable *does have* the homotopy type of CW complex.) In this case, the following holds (Chichilnisky [3, 5]):

Theorem 2. *Let f be a continuous unanimous anonymous aggregation rule which maps C^k into C. Then the homology groups of C with integer coefficients admit division by k.*

That means, that any element of any of the homology groups can be represented as k times another element (and, consequently, as a k^2, k^3 etc. multiples). Obviously, this fact restricts strongly the structure of the topological space X.

To use this topological result in our combinatorial setting, we have to derive topological data from combinatorial data: we have to use the elements of the combinatorial approach to construct a topological framework. To do it we apply the methodology relating to nerves outlined in the previous section.

First of all, we must embed the set of *profiles* X^k, into a space with nontrivial topology. To do it we embed X in the circular graph with n vertices. The set of profiles X^k is then embedded into the k-fold product of such graphs (Fig. 3). This product has the topology of (actually is homeomorphic to) a k-dimensional torus (by definition, a product of k circles!) and is composed of pieces homeomorphic to k-dimensional cubes – products of the edges of the circular graph.

This structure is not yet a simplicial complex: it has to be composed of simplices, not of cubes. To conform to the definition we triangulate the cubes (Fig. 4) so that each cube would consists of $k!$ k-dimensional simplices spanned by $(k + 1)$ vertices in X^k. All these simplices have the property that their vertices can be ordered into a linear chain, so that any two neighboring

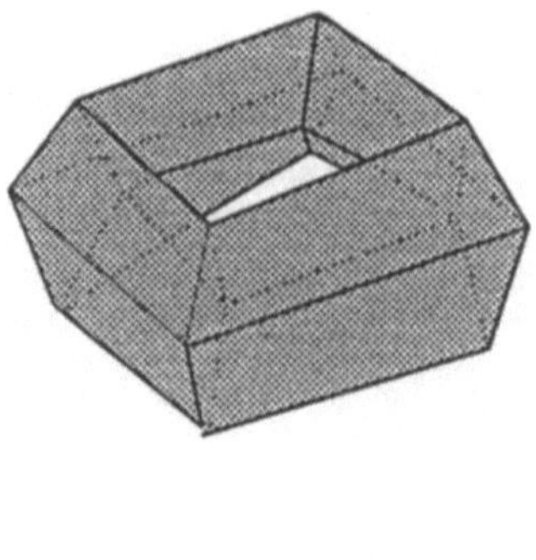

Fig. 3. Product of two circular graphs with 4 vertices each

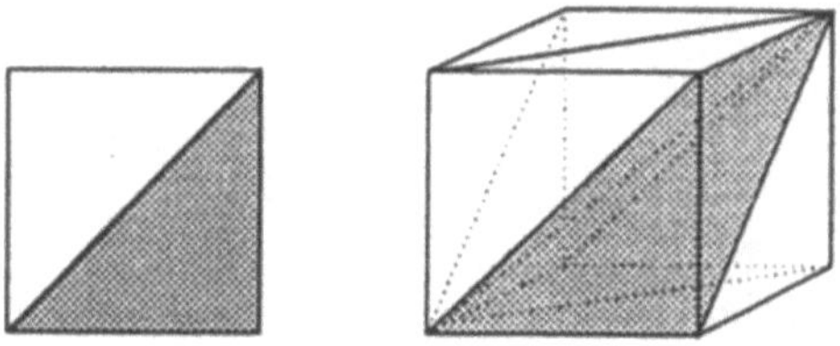

Fig. 4. Triangulations of 2- and 3-dimensional cubes. Just one of the simplices is shown

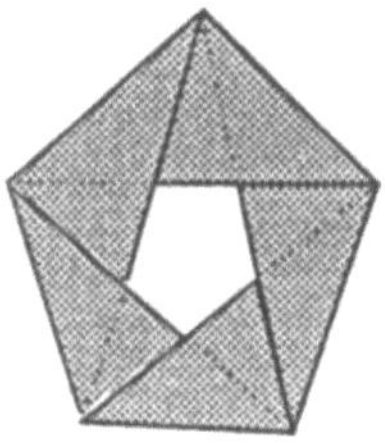

Fig. 5. The simplicial complex consists of 5 2-dimensional simplices; attached together they form an origami Moebius band

vertices in the chain, considered as k-tuples of elements in X, have all but one entry coinciding, and the two distinct elements at the remaining entry are neighboring in X. The resulting simplicial complex we will denote by N_P.

Further, we will embed X itself into a simplicial complex. One might expect it to be the circular graph itself, but this does not suffice. We construct the complex N_X as follows: we say that a simplex spanned by some vertices belongs to out simplicial complex if and only if these vertices lie in a chain of length $n + 1$ (that is consisting of n consecutive edges) in X.

The resulting simplicial complex for $n = 2$, $k = 5$ is shown in Fig. 5. It is clear that in this case it has the homotopy type of a circle (being homeomorphic actually to the Möbius strip). This fact pertains if k is large enough:

Proposition 1. *If $n > 2k$, then the simplicial complex N_X has the homotopy type of circle S^1.*

The proof is rather intuitive: the circular graph X is a subspace of N_X (considered as topological spaces). Define the projection of each of n simplices of which N_X consists to X: one can identify the $(n + 1)$ chain to which the vertices of the simplex belong with the segment $[0, n] \in \mathbf{R}$; map consecutive vertices in the chain to the consecutive integers from 0 to n and extend then the mapping to the whole simplex by linearity. One checks immediately that the mapping is well defined (here is the condition $n > 2k$ used), and that it is a homotopic equivalence.

Now we want to extend the mapping $f: X^k \to X$ sending profiles of preferred outcomes to an outcome to a mapping of the simplicial complexes $N_P \to N_X$. The mapping is already defined on all vertices of N_P and we extend it to the whole N_X linearly. This means that if a point in N_X belongs to the inferiority of a simplex (that is, it can be represented as a convex combination of its vertices), then the image of the point is the same convex combination of the images of the vertices. To have this mapping well defined one needs just to check if the image of any set of vertices spanning a simplex in N_P spans a simplex in N_X. This is obviously the case: any set of vertices spanning a simplex in N_X can be joined into a linear chain (of length n), so that any two neighboring vertices differ in at most one coordinate, and in this coordinate the elements of X are neighboring. That implies, by the Stability Axiom, that the images of the two neighboring elements of the chain are neighboring in X. Thus the image of any set of vertices spanning a simplex in N_P belongs to a linear chain in X of length at most $(n+1)$ and thus spans a simplex in N_X. This means that the image of any simplex in N_P belongs to N_X.

Let us summarize what we have earned at the stage: the usage of simplicial complexes makes it possible to move from purely combinatorial problems to topological ones, while the Stability Axiom ensures that the resulting mapping are continuous (which is equivalent to simplicial in our category of simplicial complexes). At this point one can directly apply the appropriate results from topological social choice theory and use them to yield results in the combinatorial one.

Indeed, the symmetry of the combinatorial rule $X^k \to X$ immediately implies the symmetry of the associated mapping $N_P \to N_X$. The Unanimity Axiom translates into the unanimity of the topological rule (that is the diagonal of N_P is mapped identically on N_X). Thus the Chichilnisky's theorem stated above applies, and implies that there exists no aggregation rule on choices of camping site satisfying all the axioms of stability, symmetry and unanimity. Indeed, if there were one, one could then construct as above a continuous mapping from k-dimensional torus $(S^1)^k$ into S^1 satisfying symmetry and unianimity axioms, which by Chichilnisky's' theorem tells us that all the homology groups of S^1 admit division by k – which is false! The first homology group is just the group of integers and the number 1 cannot be divided by k unless $k = 1$.

That, obviously, proves the Theorem.

3. Arrow's impossibility theorem and another topological choice theory

The model of the previous section was clearly somewhat special. In this section I want to demonstrate that Arrow's Impossibility Theorem itself can be derived from another topological choice theory, again by Chichilnisky [4].

Arrow's theory starts with the following setup. Let a natural number k be the number of voters and the natural $n \geq 3$ be the number of alternatives. The following axioms are assumed:

Independence of infeasible alternatives. The result of comparing of a pair $1 \leq i, j \leq k$ according to the aggregated preference depends only on such results in every weak order of the preference profile given.

Unanimity. If all the voters have the same preference on a pair $\{i, j\}$, then the aggregated rule has to coincide with them on the pair.

To construct a topological spaces from these ingredients we use the independence of infeasible alternatives as a cornerstone. Indeed, this axioms suggests that we do not work with the preferences themselves, but with the subsets of them consisting of preferences having given restrictions on a pair of alternatives.

Let X be the set of strict orderings of n alternatives. We deviate here a bit from the classical setting dealing with weak orders, but it does not lead to a loss of generality. Let $\sigma \in \{+, -\}$; $i, j \in [n]$, $i < j$. Sets of orders with given ranking of elements i, j we will denote as $\mathscr{U}_{ij}^{\sigma}$:

$$\mathscr{U}_{ij}^{+} = \{w \in X: i >_w j\}; \quad \mathscr{U}_{ij}^{-} = \{w \in X: i <_w j\},$$

where $>_w$ is the preference relation defined by the ranking w. Obviously, these sets $\mathscr{U}_{ij}^{\sigma}$ cover X.

What we are doing here is converting the space of orderings of n alternatives into a simplicial complex. We do this by constructing a cover of the set of orderings. The covering subset of the cover consists of set or orderings which rank a particular pair in a particular way. From this cover we can form the nerve, a simplicial complex as discussed in Section 1.

This construction is in a way miraculous: from a discrete set of n points having "no topology", we produce a simplicial complex with nontrivial topology reflecting interlacing of subsets with different rankings of pairs of alternatives, so that this very nontriviality will appear to be responsible for the impossibility paradox!

Let N_X be the nerve of the covering of X by the sets $\mathscr{U}_{ij}^{\sigma}$. The structure of N_X as topological space is given by the following

Theorem 3. *The simplicial complex N_X is homotopy equivalent to the $(n-2)$-dimensional sphere S^{n-2}.*

On Fig. 6 we show the simplicial complex N_X for X of cardinality three; it is apparent that it is homeomorphic to a circular band.

To produce a topological space from the set of *profiles*, that is k-tuples of elements of X, we use similar subsets: for any pair of alternatives i, j we fix a k-vector of pluses and minuses and consider sets of profiles in which each voter ranks i, j accordingly to the sign staying on the corresponding place. Formally, let $\bar{\sigma} = (\sigma_1, \ldots, \sigma_k)$ be a vector of sings, $\sigma_1 \in \{+, -\}$. We set

$$\mathscr{U}_{ij}^{\bar{\sigma}} = \{p = (w_1, \ldots, w_k) \in P_0: w_l \in \mathscr{U}_{ij}^{\sigma_l}\},$$

that is a profile belongs to $\{\mathscr{U}_{ij}^{\bar{\sigma}}\}$ if the l-th voter's preference in the tuple forming the profile ranges the pair of alternatives i, j in accordance with the sign σ_l.

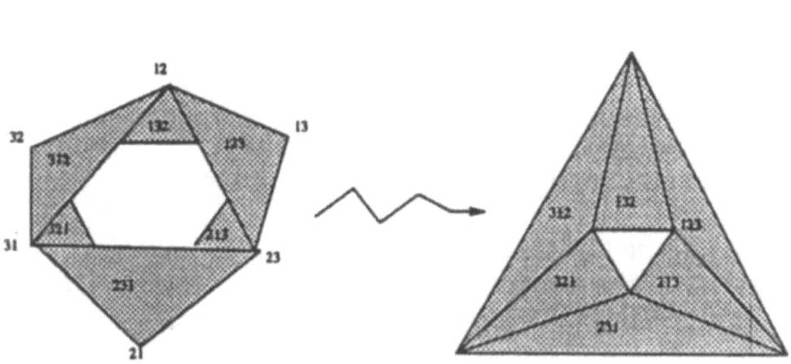

Fig. 6. Simplicial compelx N_X for orderings on three points. Vertices correspond to orderings on pairs; simplices to strict orders on X. The right complex is homeomorphic to N_X and is obviously a circular strip

Clearly, $\{\mathcal{U}_{ij}^{\bar{\sigma}}\}$ form a cover of $P = X^k$. We denote the nerve of the covering of the set of profiles P by N_P. This simplicial complex has quite a complicated structure. However, one can restrict attention only to simplices up to a certain dimensions; the homology groups in dimension d do not depend on simplices of dimensions $(d + 2)$ and higher. As we will see, the topological theory we are going to use relies only on homologies of the profile space which have the dimension of the homologies of the preference space. In our case, the space X is an $(n - 2)$-dimensional sphere, so we will be interested only in homologies of N_P in dimensions up to $(n - 2)$. Here everything is very simple:

Proposition 2. *The homology groups of N_P are zeros in positive dimensions up to $n - 2$, and $H^{n-2}(N_P) = Z^k$.*

In other words, the homologies of N_P is small dimensions coincide with homologies of the direct product of $k(n - 2)$-dimensional spheres.

Now we consider how the aggregation rule itself is represented in our reformulation of Arrow's theory. The axiom of independence says, in fact, nothing but the following: if a profile in the subset $\mathcal{U}_{ij}$ is mapped to an order in $\mathcal{U}_{ij}^{\sigma}$, then all the orders in the former subset are mapped into the latter one. Thus one can unambiguously define the mapping on the vertices of N_P into the vertices of N_X. This mapping can be extended to the mapping of the simplicial complexes by linearity. The only thing one has to check is the following: if some vertices in N_P span a simplex, then their images in N_X also do. This is quite obvious: the simplices in N_P span a simplex in N_P if and only if there exists a profile belonging to all the subsets. The image of the profile under the aggregation rule belongs to the images of the subsets. Thus their intersection is nonempty and they span a simplex in N_X.

This proves that any aggregation rule satisfying the Independence Axiom generates a simplicial mapping from N_P into N_X. We will now consider how the other axioms translates into this topological framework. Notice that the simplicial complex N_X can be embedded into N_P: one just sends $\mathcal{U}_{ij}^{\sigma}$ to $\mathcal{U}_{ij}^{\bar{\sigma}}$ with $\bar{\sigma} = (\sigma, \ldots, \sigma)$ and extends the mapping to simplices by linearity. The Unanimity Axiom says that this embedding composed with the aggregation rule is the identical mapping on the set X.

Now we return to the original topological theory of Chichilnisky. In this theory, the preferences form a topological space. For simplicity we *assume* that it is the space of linear functions on Euclidean space of norm 1; in fact one can *show* that the space of smooth preferences in Debreu's sense of Euclidean space retracts to this finite dimensional subspace of linear nontrivial preferences. These normed linear preferences form the unit sphere. So a preference for this instant is just a point on the unit sphere and a profile is just a k-tuple of such points.

The aggregation rule in this situation is a continuous mapping from the space of profiles $P = (S^d)^k$ to the space of preferences $X = S^d$, where d is the dimension of the sphere. Chichilnisky imposes the following axiom on the mapping f:

If for a couple of vectors of the Euclidean space all the preferences in a profile estimate first vector higher than second, then the aggregated preference should do the same.

This assumption is very simple and natural (as it is a counterpart of the binary unanimity axiom of the classical combinatorial theory), yet it yields a number of important consequences. Firstly, the composition of the diagonal embedding Δ: $X \to P$: $\Delta(x) = (x, \ldots, x)$ with the aggregation rule gives the identity mapping. Secondly, it can be shown that each of k voters either is a manipulator or has no influence. A manipulator is a voter with the following power. Given statements of preferences by the other agents, there is a preference he or she can state (which may not be a true preference) which will ensure *any* prespecified desired outcome (see also discussion of this concept in the papers by Heal and Koshevoy in this volume).

Clearly, a dictator is a manipulator but not inversely. Further, a voter has no influence, if under some circumstances he cannot reach a prespecified outcome of the aggregation rule, regardless of what preferences he claims.

The Chichilnisky theory states that for the topological model described above we can determine at the homological level whether a voter is a manipulator or neglectible. That means that it suffices to trace how the mapping is represented on the homological level. It can be shown that the binary unanimity (Pareto) axiom implies that on the homological level each voter is either dictator (and in this case he is actually a manipulator) or a dummy voter (and in this case he has no influence). In terms of homology groups, it can be represented by the following diagram of homomorphisms of free abelian groups:

Main Diagram

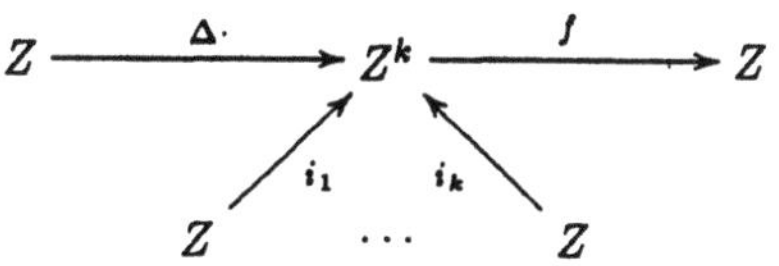

Here Δ is the diagonal embedding $n \mapsto (n, \ldots, n)$; i_l are coordinate embeddings $n \mapsto (0, \ldots, n^{l\text{th}} \ldots, 0)$ and f is the action of the aggregation rule in homologies, which is always a linear function (with integer coefficients):

$$f\colon (n_1, \ldots, n_k) \mapsto \sum_{l=1 \ldots k} a_l n_l; a_l \in Z.$$

Voter l is a dictator on the homological level if and only if $a_l = 1$; otherwise a_l should vanish and in this case the voter has no influence. The condition that the composition of the diagonal embedding and the aggregation rule is identity transforms then as

$$n \mapsto \sum_{l=1 \ldots k} a_l n = n,$$

i.e. $\sum_{l=1 \ldots k} a_l = 1$.

Obviously, that proves that exactly one of the voters is a dictator on homological level, and thus we have:

Theorem 4. *For any continuous aggregation rule on S^d satisfying Pareto condition there exists exactly one manipulator, and the other voters are neglectible.*

This is the topological counterpart of the Arrow's theory. Now we continue the reduction of the original combinatorial theory to homological calculations. We already have the main ingredient – the Main Diagram, with the middle abelian group being the $(n-2)$-nd homology group of N_P, and fringe groups being the $(n-2)$-nd homology group of N_X. The Unanimity Axiom implies that the composite mapping $f \circ \Delta$ is the identity, and what remains is to look into the meaning of the coefficients a_l. This can be done:

Proposition 3 ([2], Proposition 7). *The coefficients a_l is 1 if l is a dictator and* 0 *otherwise.*

Now we have arrived at exactly the same stage as in the case of topological theory of Chichilnisky, and the same easy calculations show that there always exists a unique dictator, thus implying the Arrow Impossibility Theorem.

4. Conclusion

In the previous sections we have seen how topological machinery works in the context of discrete problems of social choice. I hope it demonstrates clearly the possibility of using these tools in the area. However, one certainly needs to ask – why one should to undertake such circuitous path to prove (or reprove) these theorems. This issue I would like to address here.

There are two basic reasons. One of them is fairly corporative – the topology we are exploiting is in fact very geometrical and thus appeals to our intuition much better than dry combinatorial constructions. It is as with the fixed point theorem – once it is proved, it becomes a part of the arsenal and it is but natural to use it whenever a possibility arises. In fact I started the whole enterprise of reproving the Arrow impossibility theorem only because I didn't have a feeling for the proof of the theorem. On the other hand, the result is so intrinsically beautiful that I couldn't resist the desire to work out a proof which would match the result in its aesthetic appeal. It was a deep surprise to have the idea realized. Such a thing cannot be a sheer coincidence and this last circumstance suggested the second reason.

The similarities between the two theories, the classical and topological ones, are somewhat more extended than one would expect. The details seem to fit too well to represent just an analogy. I would conjecture that the homological way of proving results in both theories is a 'true' one because of its uniformity and thus can lead to much deeper understanding of the structure of social choice. To understand this structure better we need a much more evolved collection of examples of unifying these two theories and I hope this can and will be done.

References

1. Arrow K J (1950) A difficulty in the concept of social welfare. J Polit Econ 58: 328–346
2. Baryshnikov Y (1994) Unifying Impossibility Theorems: A Topological Approach. Adv Appl Math 14: 404–415

3. Chichilnisky G (1980) Social choice and the topology of spaces of preferences. Adv Math 37: 165–176
4. Chichilnisky G (1982) The topological equivalence of the Pareto condition and the existence of a dictator. J Math Econ 9: 223–233
5. Chichilnisky G, Hela G M (1983) Necessary and sufficient conditions for a resolution of the social choice paradox. J Econ Theory 31: 68–87
6. Chichilnisky G (1993) Intersecting Familied of Sets and the Topology of Cones in Economics. Bulletin AMS, 29: 2–31
7. Chichilnisky G (1994) Social diversity, arbitrage and gains from trade: a unified perspective on resource allocation. Am Econ Rev Papers and Proceedings
8. Chichilnisky G (1995) A unified perspective on resource allocation: limited arbitrage is necessary and sufficient for existence of an equilibrium, the core and social choice. In: Arrow KJ, Sen AK, Suzumura K (eds) Social Chocie Reexamined. McMillan
9. Debreu G (1971) Smooth preference. Econometrica 40: 603–617
10. Dierker E (1982) Regular economies. In: Handbook of mathematical economics (Vol II) 1: 795–830, North Holland
11. Fishburn P (1973) The Theory of Social Choice. Princeton University Press
12. Saari D (1991) Calculus and extensions of Arrow's theorem. J Math Econ 20: 271–306

Informational geometry of social choice

Donald G. Saari

Department of Mathematics, Northwestern University, Evanston, IL 60208-2730, USA

Received: 15 December 1993/Accepted: 22 April 1996

Abstract. Elementary geometry is used to understand, extend and resolve basic informational difficulties in choice theory. This includes axiomatic conclusions such as Arrow's Theorem, Chichilnisky's dictator, and the Gibbard–Satterthwaite result. In this manner new results about positional voting methods are outlined, and difficulties with axiomatic approach are discussed. A topological result about "dictatorial" behavior is offered.

1. Introduction

Much of mathematics, as we know it today, reflects its close intellectual connections with the physical and biological sciences. While it is premature to venture whether the social sciences will enjoy a similar symbiotic relationship, they do generate novel mathematical challenges. Using choice theory, I describe why I believe that responses to these challenges must reflect the greater need of the social sciences to conquer what Richard Bellman called "the curse of dimension".

The source of the curse is obvious; the social sciences rely upon the infinite dimensional information generated by "Who knows what?" "Who wants what?" and "Who is saying what to whom?" To coordinate the conflicting information coming from competing agents, we invent complicated aggregation rules, e.g., the price and other allocation mechanisms, legislative and choice rules, etc. This suggests that to analyze these procedures we should directly confront the source of the curse – we should examine the higher dimensional geometry of this information.

This is a written, expanded version of my comments at a 1993 Columbia University conference hosted by G. Chichilnisky and G. Heal; I thank both for their hospitality. Also, I thank Roko Aliprantis, Don Brown, Geoff Heal, Joe Ostroy, and Maurice Salles for comments. This research was supported, in part, by NSF IRI 9103184.

How can this be done? While there may be many ways, my research preference exploits the fact that, because the informational properties of a given mapping $F: \Omega \rightarrow \mathscr{A}$ (representing a goal, procedure, etc.) are embedded in the F level sets, it is the geometry of these level sets that reveals important features about procedures; e.g., we must anticipate that all basic results, including Arrow's Theorem, the Gibbard–Satterthwaite Theorem, and the various paradoxes that plague voting methods, are consequences of the geometric properties of information. Even more: we must expect this geometry to indicate how new methods and procedures can be designed.

A goal of this paper is to outline the surprising amount of information waiting to be extracted from this level set geometry. For instance, in Sect. 2, I show how a wide selection of issues can be uncovered just by seeking choice theory interpretations for obvious geometric facts. The ones I use are: (a) The full level set, not just a small portion, needs to be considered. (b) Different procedures must define different level sets; e.g., the level sets passing through a fixed profile can change with the procedure. (c) It is possible for the level set geometry to change radically with changes in basic parameters; e.g., the number of voters or candidates. Then, in Sect. 3, I show how geometric properties as elementary as the orientation of surfaces have deep consequences for choice theory. Recognizing the importance of neutrality, in Sect. 4 I show how it defines level sets for positional voting methods. Then, I indicate why most results about these methods is embedded in this geometry.

Since my goal is to underscore the gains available from geometry, I emphasize intuition and general ideas at the expense of details. Applications to specific problems, proofs, missing details and assumptions, and extensions either can be supplied by the reader or are found elsewhere.

2. Axiomatic approaches

We have become accustomed to the muscle power of the axiomatic approach as illustrated by Arrow's Theorem (1963) showing that the set of procedures satisfying some seemingly innocuous requirements is empty, and the Gibbard (1973)–Satterthwaite (1975) Theorem establishing that strategy proof mechanisms are dictatorial. Examples of positive assertions include the axiomatic characterization of "best choice" procedures (Sen 1977), the Borda Count (Young 1974; Saari 1990), and conclusions identifying "not the unique best" methods (Baigent and Gaertner 1993). Because axioms are so widely accepted, I selected them to illustrate that by seeking choice theory interpretations for the obvious geometric facts (a–c), we can enrich our understanding of the issues.

2.1. Point a – what else happens

Point a cautions us not to emphasize special profiles of a level set at the expense of others. As I show, a choice theory interpretation is that we must worry whether an assertion may, unintentionally, mislead us. To do this, notice that the axiomatic approach is a true success story, in part, because it helps resolve a serious difficulty. Namely, the real problem in choice theory is not a lack of procedures, but an overabundance of them. To handle this problem, we consider only methods enjoying properties we want satisfied; the axiomatic approach identifies them.

Some axioms, such as neutrality and anonymity, capture a sense of fairness. *Anonymity*, for instance, ensures that no voter can unduly influence the conclusion because each voter has the same impact. Similarly, *neutrality* requires the candidates to be treated equally; if their names are interchanged, so are the results. If Martha needs 51% of the vote to be elected, so does Ruth. A related fairness axiom requires a candidate strongly supported by the voters to be selected. Conversely, a candidate highly resented by most voters, say almost two-thirds or more of them, should be rejected. The following quantifies these expectations.

Definition. A choice procedure is *majority preserving* if, when a profile $\mathbf{p}$ has more than half of the voters with c_j top-ranked, then $f(\mathbf{p}) = \{c_j\}$. A choice function is *minimally respectable* if, when a profile $\mathbf{p}$ has c_j bottom-ranked by at least [one less than two-thirds] of all voters, $c_j \notin g(\mathbf{p})$. □

With $n = 3$ candidates, it is natural to admit only procedures satisfying $A_1 = \{$*anonymity, neutrality, and majority preserving*$\}$. The only positional method (where points are assigned to first, second, and third place candidates) satisfying $\mathscr{A}_1$ is the plurality vote. (See, for example, (Saari 1994b, p. 345).) This constitutes a strong argument for its use. While this is the good news, "a" cautions us to worry about what else can happen. To do so, recall that axioms also help eliminate procedures. For instance, an important reason to identify those procedures that satisfy $\mathscr{A}_2 = \{$*anonymity, neutrality, and can violate minimal respectability*$\}$ is to avoid them. Again, the plurality method is the only positional method in this class! (Saari 1994b, p. 345). This assertion creates an excellent reason to shun the previously lauded procedure.[1]

The source of this $\mathscr{A}_1 - \mathscr{A}_2$ conflict is clear; axioms emphasize specified properties (e.g., $\mathscr{A}_1$) while ignoring others (e.g., $\mathscr{A}_2$), so they emphasize selected profiles from level sets while ignoring potentially important consequences generated by the other profiles. As this example illustrates, the ignored properties may provide reason to doubt the merits of the procedure. In fact, because all procedures have strengths and weaknesses, it is not difficult to invent desirable sounding axioms to promote any specified procedure, or to construct an alternative axiom set that appears to disqualify the same method! This is possible simply by creating nice sounding axioms that emphasize a particular section of the level sets at the exclusion of others. Thus, a lesson learned from "a," which runs counter to a tacit theme in the literature, is that axiomatic conclusions must be treated with skepticism and care. More generally, when comparing procedures, we must compare them over all profiles.

---

[1] For $n \geq 3$ candidates, replace $\frac{2}{3}$ with $(n-1)/n$. Thus, while in $n = 10$ candidate elections, only the plurality vote can guarantee the selection of a candidate top-ranked by over half of all voters, it also is the only positional method where a candidate could be selected even though one less than 90% of all voters have her bottom-ranked!

2.2. Point b – changing structures

Point b suggests comparing how the outcomes associated with a profile can change with the choice of a procedure. By applying this to $\mathscr{A}_1$, we discover the uncomfortable fact that while the axiomatic approach can anoint certain procedures as "the best" thanks to rare, isolated profiles (point a), a careful examination of these profiles can disclose that even here the touted outcome is the wrong one! Thus, axiomatic conclusions can be seriously misleading.

To illustrate with $\mathscr{A}_1$, we first need to identify all profiles where only the plurality method selects the majority candidate. (Actually, none exist.) Indicative of all such profiles is where 50 000 voters have the preference Joyce $\succ$ Lillian $\succ$ Connie and 50 000 have Lillian $\succ$ Connie $\succ$ Joyce. Clearly, Lillian is the robust natural choice. (Half have her top-ranked; the rest have her middle-ranked. No other candidate commands such favored support.) This sensible conclusion is supported by all positional methods *except the plurality vote* which has Joyce tied with Lillian. By adding another Joyce supporter, Joyce becomes the Condorcet and plurality winner – yet almost all other positional methods correctly select Lillian.[2] So, by examining this profile and how it changes level sets (i.e., outcomes) with changes in procedures, we discover that $\mathscr{A}_1$ actually discloses a flaw, not a virtue, of the plurality vote.[3] Thus, while the assertion that *only* the plurality method is "majority preserving" seems attractive in the abstract, it loses all luster when the supporting profiles are analyzed. (As an exercise, the reader can find fault with the fact that the antiplurality vote is the only positional method to ensure that a candidate bottom-ranked by over half of the voters is not selected.)

2.3. Point c; comparing models

Issue c cautions that by varying basic parameters the structure of level sets (and conclusions) can seriously change. To examine what this means for choice theory, I use the extreme setting where the number of issues or voters goes to infinity. Here, "c" requires determining whether the level set

[2] Lillian wins with procedures $(1, s, 0)$ where $s > 1/100\,000$ is the number of points assigned to a voter's second ranked candidate. To extend the conclusion to, say, $s > 10^{-9}$, replace each 50 000 with 500 000 000 voters. By using the procedure line and profile representation from (Saari 1994b), it follows that all profiles illustrating that only the plurality method selects a majority winner start with a lower dimensional (hence, rare) set of profiles formed by a convex combination of p_1 where half of the voters have $J \succ L \succ C$ and half have $L \succ C \succ J$, p_2 where half have $J \succ L \succ C$ and half have $L \succ J \succ C$, and p_3 where half have $J \succ L \succ L$, and half have $L \succ C \succ J$. If p_1 consists of several voters, then L wins with all positional procedures except the plurality vote; here $J \sim L$. Adding a voter with J top-ranked breaks the tie. To ensure that only the plurality vote has this property, the number of voters involved in p_1 must approach infinity; thus, the essence of the example dominates any such profile.

[3] The plurality vote ignores all information about voters lower ranked candidates, so it, along with the Condorcet winner, is incapable of discerning situations where a majority candidate is not the natural choice. In fact, serious doubt can be cast on both procedures just by expressing this fact in axiomatic form.

geometries of models involving infinite and finite numbers are related; i.e., do conclusions from one choice model tell us anything about the other? If not, then we may be studying nonexistent issues; i.e., issues that cannot be supported by any real profile.

A way to explain my point, is to verify that $\sum_{j=0}^{\infty}(\frac{1}{2})^j = 2$. While elementary, *nobody* could ever carry out the summation! (If someone tried, an infinite number of terms always would remain.) The difficulty is that this is not a real summation; it only indicates that the value of $\sum_{j=0}^{N} 2^{-j}$ is arbitrarily close to 2 once N is sufficiently large. Namely, the real purpose of infinite summations is to identify when a sense of invariance and constancy prevails for large *finite* values of N. To conveniently discover these invariances, a fictitious infinite summation is invented (it is a *limit*, not a summation) and supplemented with computational techniques.

Supposedly, this philosophy justifies models with an infinite number of agents, commodities, candidates, etc. After all, not even a mathematician living in abstract worlds would care about a fictitious infinite society; these models are mere conveniences invented to discover properties enjoyed by models with large numbers of voters or issues. So, while problems about infinite models present amusing intellectual challenges, they have relevancy only should they uncover real properties about *finite* models.[4] In geometric terms, we need to understand how the level set geometry in finite spaces is mimicked by that in the infinite space.

Just raising this issue identifies the problem; commonly used mathematical constructs involving infinities, such as Banach and Hilbert spaces, measure theory, etc., admit behavior forbidden by "large, finite" models! (So, a conclusion based on such behavior may define a nonexistent issue; i.e., it corresponds to an empty set of *real profiles.*) Namely, conclusions from infinite models only identify "candidate assertions;" the critical next step is to determine which (if any) represent valid conclusions about large finite models. This verification step is similar to what happens for a calculus maxima problem. Critical points identify potential answers; the next step is to determine which candidate outcomes correspond to valid maxima. Indeed, the calculus problem is unsolved until this step is completed. Similarly, *infinite model conclusions only identify candidates for real assertions; the critical next step is to prove which ones, if any, represent limiting behavior of finite models*! Unfortunately, this critical verification step, which is needed to justify these models and which indicates how level sets for finite models compare with the infinite model, usually is ignored.

This is a real, not just a technical concern. After all, although Arrow's impossibility conclusion holds for any number of agents, there exist nondictatorial methods (Fishburn 1970; Hansson 1976) for infinite number of agents. Had only the infinite agent setting been proved without connecting it to the finite case, we might believe that Arrow's Theorem fails with enough

[4] For instance, the earlier limit argument showing why only the plurality method satisfies $\mathscr{A}_1$ is a choice theory example of a constancy property; the argument shows how to design a real (finite) profile to demonstrate the assertion for each non-plurality positional method.

voters. What a serious error this would have been! Similarly, it is not difficult to find in the literature other infinite model conclusions (from economics, choice theory, etc.) which must be treated as extraneous solutions (reflecting the chosen mathematical technique) as they fail to be indicative of what happens in finite models. The reason the verification step is mandatory for social choice and economic models but not for mathematical summations, is that infinite summations are *defined* as the limit of large finite settings. While this may be the intent, if it is not part of formal definition or technique[5] of infinite models in the social sciences, then the tacit limit assumption – the content of "c" – must be verified. More bluntly, which outcomes are supported by real profiles?

Because infinite models find their justification from the limit theorems, this introduces flexibility in designing axioms for a fictitious infinite model. After all, worrying about a "fair" definition for anonymity, neutrality, etc. for such an infinite, pretend society is the modern equivalent of worrying about how many angels can dance on a pin head; the answer doesn't matter.[6] The real issue is whether a definition created for an infinite model survives the verification step. Thus, as "c" requires, the emphasis must be on the limiting structure of the level sets for finite models. To illustrate, a definition of neutrality need not require *all* permutations of the candidates' names to be respected; it could allow all permutations of *any finite* subset. Both definitions are equivalent with finite numbers of candidates, but the difference is mathematically significant in the limit! In fact, because definitions of neutrality (anonymity, etc.) using terms such as "any set" implicitly define a limit process, we must anticipate them to introduce extraneous, misleading conclusions. Examples are easy to find and design; they occur with those published results where the infinite model behaves differently than any finite model. To design new examples, create situations where the order of taking the limits matters.

3. Refined results through orientation of surfaces

I now demonstrate how elementary properties of level get geometry, such as the orientation of surfaces, provides a rich supply of new answers, extensions, and insight into troubling assertions from this field. To illustrate, I selected the geometry associated with the commonly used "dictatorial" assumptions. But first, we need some notation.

Let Ω^j represent the characteristics of the jth agent we want to model; usually $\prod_{j=1}^{a} \Omega^j$ is the profile space. If $\mathscr{U}$ is the outcome space and P is the procedure for a agents, then the outcome[7] is represented by

$$P : \prod_{j=1}^{a} \Omega^j \to \mathscr{U} . \tag{3.1}$$

[5] So, a positive feature of nonstandard analysis applied to social models, as pioneered by Don Brown and now being used by A. Khan and his group, is that the limit process is built into it.

[6] Yet, as manifested by the lively debate in this conference, arguments about the appropriateness of axioms typically emphasize "fairness" implications for the infinite, fictitious model; instead, they should focus on the limit process.

[7] Other papers in this collection use the notation $f: P^k \to P$.

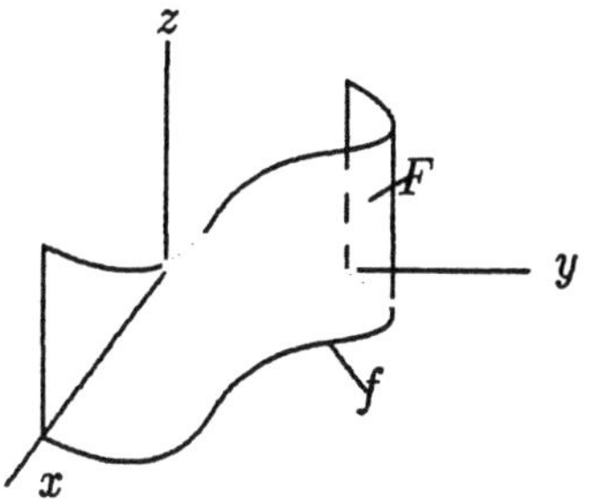

Fig. 1. Level sets

A level set identifies all ways a particular outcome can occur. The collection of level sets of P partitions the *profile space* $\prod_{j=1}^{a} \Omega^j$, and the approach being promoted here is to analyze P through the geometry and relative positions of these surfaces. This partitioning forms a *foliation.*

3.1. Geometry of dictators

To develop intuition for choice theory interpretations of the orientation of sets, recall that, in general, a level set of a function $f(x, y)$ is a curve in the x–y plane. By extending f's domain to R^3 – represented by $F(x, y, z) = f(x, y)$ – the level sets are two-dimensional surfaces parallel to the z axis.[8] See Fig. 1. This parallel orientation reflects, and is equivalent to, F's lack of dependency upon z. Similarly, as $g(z)$ is a function of one variable, its R^3 level sets are planes parallel to the x–y coordinate plane. Conversely, if all level sets of a function are parallel to the x–y plane, that function is determined solely by z values.

To show how "parallel orientations" arise in choice theory, let $\prod_{j=1}^{3} \Omega^j \subset R^3$ where each axis of R^3 identifies the characteristics of a particular agent; e.g., in Fig. 2 each point of Ω^j on the jth axis identifies a possible choice for the jth voter's top-ranked candidate. If the P level sets are in planes parallel to the x–y coordinate plane, then (according to the above discussion) P ignores the preferences of the first two voters – the P outcome is strictly determined by one agent's preferences ("Don" in Fig. 2a). Because this parallel orientation requires each level set to include *all* characteristics of the first two agents (Carl and Gene) but just the single point characterizing Don's choice (the z value), each P level set is uniquely identified by its sole z value. As this requires the P outcome to be uniquely determined by Don's preferences, P is a dictatorship, an antidictatorship, or some other procedure where the outcome is uniquely determined by Don's characteristics.

This "dictator" conclusion only requires each P level set to be completely identified, or indexed, by the characteristics of a single agent. Whenever this occurs, call the procedure "dictator-like;" the outcome is strictly determined by the characteristics of this agent. This definition admits considerable flexibility in the associated orientation of level sets for discrete models; e.g., by exploiting the distance between alternatives, the orientations of level sets can

[8] To obtain the F level sets, draw a line parallel to the z axis through each point on the f level set; mathematically, each F level set is the product of a f level set with the z axis.

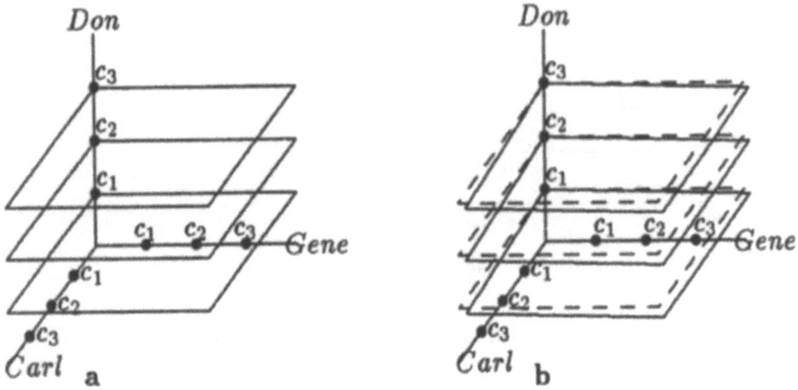

Fig. 2. Dictators and de facto dictators

be tilted away from the "dictatorial parallel" while still satisfying the dictator-like conditions. This is illustrated in Fig. 2b where, although the level sets are not parallel to the x–y plane, each level set is completely identified by the z value.

To find a choice theory interpretation for the tilted surfaces, consider an election where ballots are tallied with $(1, s, 0)$ points (i.e., one point to a top-ranked candidate and $s \in (0, 1)$ to a second-ranked one). Suppose after each of the k voters casts a single ballot, my ballot counts as though it were cast by $1 + k/2$ voters. With this biased procedure, while a voter might believe he influences the outcome, he doesn't – I am the de facto dictator. (The $0 < s < 1$ restriction ensures that IIA and Arrow's other conditions are satisfied.) As there are three possible choices for top-ranked candidate, this procedure has three level sets. If the procedure had ignored the other voters, the level sets would be dictatorial parallel to those axes representing the ignored voters' preferences. But with this procedure, representing the slight impact their votes have on the finally tally (but not on the final outcome) is the slight tilt of the level sets.

Similarly, in Fig. 2b, if values along each axis correspond to the total number points a candidate receives in this highly biased election, where the level sets correspond to the winning candidate, then the different orientations may manifest, for example, weighted voting procedures where the weight of the votes of the three voters are given by $\omega_j \geq 0,\ j = 1, 2, 3$; thus the vote of the jth voter counts as though ω_j voters voted. (So, a level set identifies the "winner," the tilt corresponds to the number of points she receives corresponding to each possible profile; it is determined by the ω_j values.) As long as

$$\omega_3 > \omega_1 + \omega_2, \tag{3.2}$$

the third voter is a "de facto dictator;" any thought that the other voters have an impact is a delusion. (Notice how the sequential dictator follows with a further assumption that, say, $\omega_3 > \omega_2 > \omega_1$.)

Observe the many different level set orientations corresponding to procedures ranging from a dictator to the various de factor dictators. These procedures are related because the different orientations of level sets define a continuum of de facto dictators that can be continuously deformed into an Arrow dictator.[9] So, while Arrow's Theorem mandates a "dictator," the level set geometry proves that his conclusion is supported by a wealth of procedures. Even more, by relaxing the restrictions on parameter values defining these de facto dictatorships, non-dictatorial methods arise. This happens, for instance, when Eq. (3.2) is violated, or when $\omega_1 = \omega_2 = \omega_3$ which defines the plurality vote. In other words, we have established that there exist continuous connections between Arrow's dictator and methods in actual use.

3.2. *Positional voting paradoxes*

Another way to illustrate the importance of orientation is with positional voting. A positional voting method for $n \geqslant 2$ candidates is defined by a *voting vector* $\mathbf{w}^n = (w_1, w_2, \ldots, w_n = 0)$ where $w_i \geq w_{i+1}$, $i = 1, \ldots, i - 1$, and $w_1 > 0$. Here, w_j points are assigned to a voter's jth ranked alternative and each candidate is ranked according to the sum of points she receives. Clearly, each voting vector $\mathbf{w}^n$ determines a unique orientation for the level sets of profiles defined by each tally. Different orientations determine different properties, so different choices of $\mathbf{w}^n$ generate different properties.[10]

As I show next, an intuitive explanation for the many perplexing voting paradoxes involves nothing more complicated than an easy extension of the obvious fact that all level sets of a function depending on all three variables, say $F(x, y, z) = x + y + z$, cannot be parallel to any axis; if they were, it would contradict F's dependency on that variable. Conversely, choose any function $G(x, y)$; because it does not depend on z, its level sets must be parallel to the z axis. Thus, the *orientations of the G and F level sets differ significantly simply because they involve different sets of variables.* In particular, this difference in orientations forces the level sets for the different functions to cross one another (see Fig. 3), consequently, the G value does not determine, nor can it be determined by, the F value. This simple geometry explains the impact of missing or irrelevant alternatives!

To see why, start with $n = 3$ candidates. The critical information from a profile concerns the fraction of all voters that rank each candidate in first and in second position. Represent this information with a point in R^6 where each candidate is assigned two coordinates; the jth one provides the fraction of voters that have her jth ranked, $j = 1, 2$. Now, a specified $\mathbf{w}^3$ tally must involve coordinates from all three candidates, and, with the exception of the plurality vote, it must involve values coming from all six coordinates.[11] This

[9] With weighted voting, this deformation is achieved by holding ω_3 fixed while letting $\omega_1, \omega_2 \to 0$. (Using other notation, we say that the de facto dictators are homotopic to a dictator.) This represents a continuum of ways to alter the orientation of the level sets while respecting the dictator-like condition.

[10] Because the level sets are in a $n!$-dimensional space, new kinds of geometric techniques are needed to extract these properties. See Sect. 4.

[11] The plurality vote uses only information from the three coordinates representing first place status for the voters. This "lost" information forms a way to explain its deficiencies.

dependency means these (affine) level sets cannot be parallel to any axis. On the other hand, when only the candidates $\{c_i, c_j\}$ are pairwise compared – no matter what the procedure – the level sets *must* be parallel to both of the axes identified with the missing candidate, so this creates a situation analogous to that displayed in Fig. 3. Just this difference in orientation of the different kinds of level sets (the three pairwise level sets and the positional level set) means that the level sets of one procedure cannot reside in the level sets of another procedure. In terms of choice outcomes, this means that *we cannot expect the pairwise ranking of a pair to conform with its relative ranking in a positional election outcome.* Furthermore, when one considers the radically different orientations for the level sets for each of the three pairs, it follows that they must intersect one another – thus cycles and other paradoxical outcomes must be expected. (All of this can be re-expressed in algebraic topological terms, but the simpler geometry is clearer).

More generally for $n > k \geq 2$, the same geometry dictates that for any positional voting method, the ranking of a subset of k candidates and their relative ranking within the positional ranking of n candidates need not have anything to do with one another! It even suggests that for a fixed profile, the positional voting rankings of $n - 1$ subsets of candidates, $\{c_1, \dots, c_n\}$, $\{c_1, \dots, c_{n-1}\}$, $\dots$, $\{c_1, c_2\}$ need not be related in any way whatsoever! This is the case. (For a characterization of happens with positional voting, see (Saari 1989, 1993, 1994b).) As another of the many possibilities, notice that level set geometry (in R^6 for three candidates, in $R^{n!}$ in general) requires the level sets for different choices of $\mathbf{w}^n$ to intersect, so the outcomes can differ. This application of "b" to positional methods is the origin of the result (Saari 1992b) showing that a single ten-candidate profile can generate over 84 million different election rankings when the choice of $\mathbf{w}^{10}$ is varied.

3.3. A return to Arrow's Theorem

A similar argument provides geometric insight into Arrow's Theorem. Because IIA requires certain candidates to be ignored with each pairwise comparison, it requires the level sets (of the unknown procedure) to be parallel

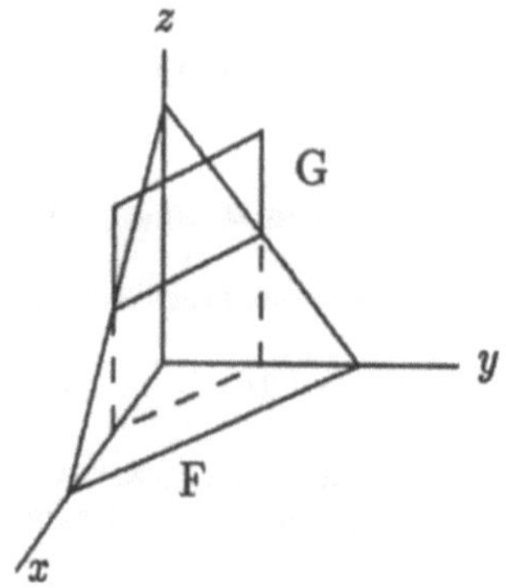

Fig. 3. Geometry for IIA and voting paradoxes

to any axis identified with the ignored candidates. The unanimity and universal domain assumptions, on the other hand, force the orientations of the level sets of the welfare (or choice) function to violate these parallel requirements. (This is because the mapping cannot ignore any candidate.) Thus, this geometric incompatibility (which is similar to Fig. 3) generates the impossibility assertion. The geometric reason we need two or more voters is that with a single voter each level set is a single point. As an "orientation" cannot be defined with a point, the necessary conflicting geometry (in the level set orientations) requires at least two agents. (For different descriptions and other consequences of IIA, see (Saari 1991b, 1994a,b).)

The geometry of Arrow's axioms, then, imposes incompatible orientations on the level sets. In fact, this incompatibility still holds even without the extreme IIA orientations. Just as above where an Arrow dictator is identified with a wealth of procedures, we can relax IIA. For example, when comparing c_1, c_2, instead of totally ignoring the relative ranking of c_3, we could attach minimal weight to c_3's positioning to measure the "intensity" of this binary comparison (see Saari 1994a, b). This intensity information tilts the orientation of the level sets away from the parallel orientations (hence it creates "relaxed IIA" conditions). While it is trivial to find appropriate conditions on these intensity conditions that preserve the geometric conflict and Arrow's assertion, there are other weight conditions on c_j's positioning that now admit possibility theorems. It is interesting that once the conflict is removed, the modified IIA conditions yield the Borda Count as the only possible positional procedure. (This is a geometric outline of (Saari 1994a).) Again, notice how this geometry provides a continuous class of IIA axioms connecting dictatorships with certain positional and other methods. Indeed, we are learning from this geometric approach that continuous classes of this type must be expected with choice theory. (The next section indicates that this omnipresent theme is related to "homotopy" assertions. A reader familiar with these algebraic techniques will have no difficulty supplying details.)

3.4. Chichilnisky dictators and a beach party

Chichilinsky's (e.g. 1982) important work and the extensions by Chichilnisky and Heal (1983) also can be understood with this geometric approach.[12] A simple version of her basic formulation identifies a person's preferences with a point on a circle S^1. This modeling is intended to represent directions of a normal vector for a certain class of utility functions. For the reader uncomfortable with this abstract formulation, replace it with the choice problem confronting a group of people choosing a picnic spot on the beach. As the beach surrounds the lake, it can be represented by a circle S^1 where the jth person's preference is determined by the point $p_j \in S^1$. With no restrictions on each person's choice of $p \in S^1$, the selection problem is

$$P: S^1 \times \dots \times S^1 \to S^1.$$

[12] My discussion differs from the other papers in this collection discussing homotopic rules in that I emphasize intuition and demonstrating other directions.

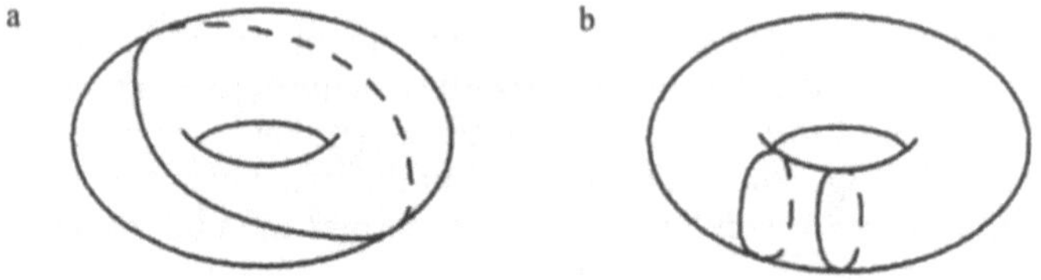

Fig. 4. **a** The unanimity line, **b** vertical dictator

Using different sets of reasonable assumptions, Chichilnisky shows that P must be "homotopic to a dictator". Her statement is falsely interpreted as identifying still another dictatorial setting.[13] Instead, the geometry of level sets discloses a wealth of interesting non-dictatorial procedures.

Again, a decision rule P is characterized by its level sets. One of Chichilnisky's principal assumptions, unanimity, anchors each P level set at a particular location. Namely, as $p \in S^1$ is the chosen beach site should everyone want that spot, the profile $(p, p, \ldots, p)$ is in the p level set. As this assumption plays an important role in forcing the conclusion, I will describe its geometry for $n = 2$ where the profile space, $S^1 \times S^1$, is a torus – the surface of a donut.[14] The unanimity assumption defines the *unanimity line* on the torus indicated in Fig. 4a – the line of points (p, p), $p \in S^1$, is where the agents agree.

To simplify the geometry, recognize that to construct a circle, we glue together the endpoints of an interval. Thus, a circle can be replaced with a line interval if we remember that its endpoints represent the same point. Similarly, to see the properties of the torus, consider a square where the first voter's preferences are identified with horizontal (x) values and the second with vertical (y) values. As the second voter's preferences come from a circle, glue the top and bottom edges together; this generates a cylinder. Similarly, identifying the extreme vertical edges for the first voter requires connecting the ends of the cylinder to create a torus. Thus, by remembering how the edges are identified, we can use the simpler square. Here, the diagonal line in Fig. 5 represents the unanimity line.[15]

So, how can level sets be defined? In the two-agent setting, each level set could be a circle; one class of circles foliates the torus horizontally while

---

[13] This was clear from conference comments; the following elaborates on my response describing why this is false.

[14] To create the torus, attach a vertical circle to each point on a horizontally positioned circle. Points on the horizontal circle represent the first agent's preferences for the beach party; for each choice, the vertical circle indicates the second agent's possible choices.

[15] Remember, level set can't cross, so each P level set meets the unanimity line in one and only one point. By requiring P to be a continuous mapping, each level set must be closed and connected. Thus level set can't elude the unanimity line restriction by introducing gaps. With the square, the "no gap" condition requires that if a line passes through the left edge, it must pass through the right edge at the same height so they will be connected after the gluing. Similarly, if it passes through the bottom edge, it also passes through the top edge at the same distance from the right side. Notice how these simple assumptions significantly restrict the possible positioning of the level sets.

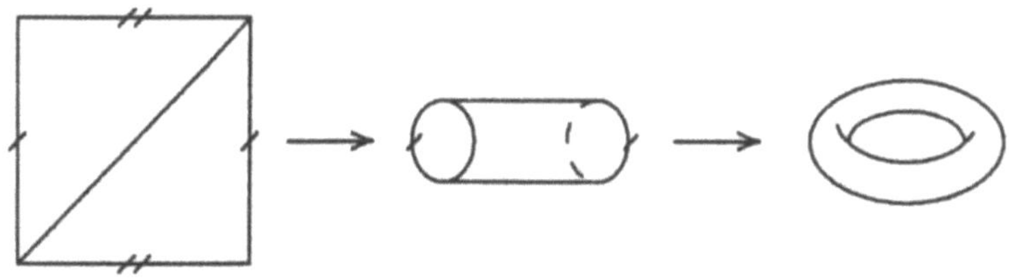

Fig. 5. Converting a square into a donut

another class slices the torus vertically as indicated in Fig. 4b. (In the square, this corresponds to horizontal or vertical lines.) Either division forces the level sets to be parallel to an agent's characteristics, so they are ignored. Thus, such a P has the beach site uniquely chosen by one person – the dictator. But, a dictator is *not* the only admissible procedure. Instead of choosing level sets that are either vertical or horizontal circles, distort the level sets so each passes through the correct point on the unanimity line but elsewhere it is off the reference dictatorial level sets (see Fig. 6). As these level sets are not dictatorial, I call then *dominant voter procedures* because one voter plays a dominant (but not total) role in determining the outcome. (A mathematical explanation is outlined below.)

To describe these new schemes with the picnic example, after the dominant person chooses the picnic site, the second person slightly refines it; maybe moves it out of the sun. These schemes, then, can be viewed as a process where after the dominant person chooses, certain other people provide (limited) refinements. As *more than one person* determines the final site, dominant voter methods are not dictatorial! Yet the domination factor (forced by the unanimity line) constrains the role of other voters; this constraint has the flavor of Eq. (3.2). These procedures, then, are continuous versions of the weighted voting schemes that define the de facto Arrow dictators; an important difference is that dominant procedures are *not* dictatorial; other voters *can influence* the outcome![16]

No matter how imaginatively one draws these level sets, it is geometrically obvious that the unanimity line severely restricts the role of the non-dominating agents. Also, as true with weighted voting, each procedure defines a continuum of procedures obtained by decreasing the weight (i.e., influence) of the non-dominant voters. Geometrically, the gaps between the reference dictatorial line and the actual level set is continuously decreased in the same way used with "dictators" and "defacto dictators" and with "IIA" and the "relaxed IIA" conditions. The mathematical term for this deformation is that the new dominant voter procedures are *homotopic to a dictator*. This is a benign interpretation of Chichilnisky's assertion.

The "homotopy" argument supports the adjective "dominant" because it is impossible to continuously deform one kind of dictator to another through

[16] To design other schemes, note that nothing forbids a level set from including a small two-dimensional region, or from crossing backwards to create a small "S." Smooth choices of P are given by all dynamical systems (vector fields) on the torus where the solutions pass through the unanimity line once.

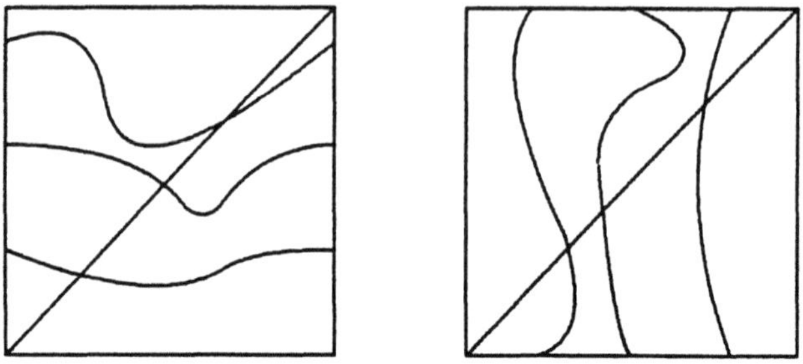

Homotopic to horizontal dictator. Homotopic to vertical dictator.

Fig. 6

these new procedures. If we could, we would be able to continuously deform horizontal lines so they all become vertical lines. To develop intuition with the square (see Fig. 7a) why this cannot be done, observe that if it could, then there would be a transition where an endpoint (say, the left one of a dashed line) of a former horizontal line passes through a vertex (say, a top one) of the square. As the endpoints of the dashed lines must have the same height, both endpoints are on the top edge. But this requires the line to pass through the unanimity line a second time, which is prohibited. For our purposes, the impossibility of transforming procedures from one partition set to another allows each class of procedures to be identified by the dictatorship it admits; this dictator is the "dominant voter" – even though it is easy to design procedures where, locally, another voter has the dominant role in the choice.[17]

3.5. Continuity is a problem

We have learned from level set geometry that "dictatorial conclusions" identify those undesirable situation where a single voter dominates (but, not necessarily dictates) the decision procedure. More generally, we want to understand which assumptions (axioms) force settings where most voters are disenfranchised in some, if not all, decisions; it doesn't matter whether this is due to a dictator or whatever. Thus the goal is to characterize conditions which force a continuous

$$F:\prod_{i=1}^{a} S^k \to S^k \tag{3.3}$$

to be identified with a mapping where only one voter has a serious say in the outcome. The outcome only up to that allocated to the non-dominant voters

[17] To do so, distort a horizontal line so that it is nearly vertical near the unanimity line. Near these positions, the "vertical" agent has the most say in the choice.

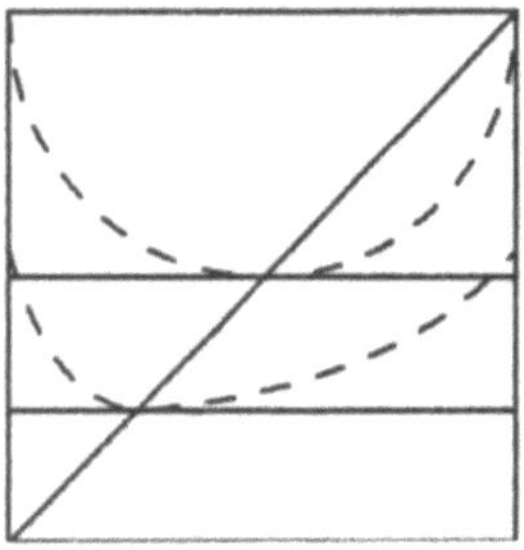

Fig. 7. Distorting an horizontal dictator

in the above schemes.) Mathematically, this means that F *is trivial over* $a-1$ *of the factors.*

With this restatement, the real problem is to determine when procedures must be (essentially) nonparticipatory over decisions. It is easy to find numerous conditions of this type. To illustrate, the following resulted from a conversation with my colleague Dan Kahn. While the assertion is immediate from a topological perspective, it has the shocking conclusion that when considering S^n where n is even, *continuity is sufficient to force the procedure to be (essentially) nonparticipatory over all decisions!*

Theorem. *For* $a \geq 2$ *agents and an even integer* $n > 0$, *let*

$$P: \prod_{i=1}^{a} S^n \to S^n$$

be a continuous mapping. The mapping P *is trivial over at least* $a-1$ *factors.*

The following short proof is intended only for those readers familiar with topological arguments.

Proof. It suffices to consider the case where $a=2$. Here, $H^n(S^n) \approx Z$ with the generator α. Similarly, it follows that $H^n(S^n \times S^n) \approx Z \oplus Z$ with generators $\alpha \otimes 1$, $1 \otimes \alpha$. By dimensionality arguments, $H^{2n}(S^n) = 0$ while $H^{2n}(S^n \times S^n) \approx Z$ with $\alpha \otimes \alpha$ as the generator.

Connecting these values is $P^*(\alpha) = k(\alpha \otimes 1) + l(1 \otimes \alpha)$ which defines the integers k, l. On the other hand, we know that $k = \deg P \mid S^n \times *$ while $l = \deg P \mid * \times S^n$.

The dimensionality statements ensure that $\alpha^2 = 0$, so $P^*(\alpha^2) = 0$, or, from the structure of P^*, that $(P^*(\alpha))^2 = P^*(\alpha^2) = 0$. It remains to compute the product

$$0 = (P^*(\alpha))^2 = [k(\alpha \otimes 1) + l(1 \otimes \alpha)] \cdot [k(\alpha \otimes 1) + l(1 \otimes \alpha)].$$

(Recall, $(\delta \otimes \gamma) \cdot (\mu \otimes \varepsilon) = (-1)^{|\gamma||\mu|}(\delta\mu \otimes \gamma\varepsilon)$.) This leads to the expression

$$k^2(\alpha^2 \otimes 1) + kl(\alpha \otimes 1)(1 \otimes \alpha) + lk(1 \otimes \alpha)(\alpha \otimes 1) + l^2(1 \otimes \alpha^2)$$

$$= 0 + kl(\alpha \otimes \alpha) + (-1)^{n^2} kl(\alpha \otimes \alpha) + 0 = 0.$$

As n is even, $(-1)^{n^2} = 1$, so $kl = 0$. The conclusion now follows. □

This conclusion, asserting that *for even values of* n *just continuity severely limits the impact of almost all voters,* extends to all $n \geq 1$ values by adding standard assumptions (such as unanimity) reflecting properties of choice theory and economics. Just choose assumptions that restrict the degree of each component to 0 or 1; then, a similar argument holds with slightly more complicated computations for $n = 1, 3, 7$.

3.6. Gibbard–Satterthwaite

To complete my illustration of the power of orientation, I turn to the important Gibbard-Satterthwaite Theorem. I view this theorem in terms of dynamics – after all, it describes the consequences of changing from sincere to strategic profiles.[18] To manipulate the procedure by altering the outcome, the strategic profile must enter a new (personally more favorable) level set. Therefore, think of this theorem in terms of a "profile motion" where, to change the outcome, the profile change passes through the boundary separating level sets. For intuition, imagine shooting a water pistol; the nozzle represents the sincere profile ($\mathbf{p}_1$) while the ejected water represents the strategic profile ($\mathbf{p}_2$). Whether we can hit a specified target (the separating boundary) depends on how we aim the pistol ($\mathbf{p}_2-\mathbf{p}_1$) and where it is ($\mathbf{p}_1$). Whether the target will be hit depends on whether we can aim it in the same general direction as a vector perpendicular to the target (that points toward the target). The goal is to position the target (i.e, design a procedure) so that no matter where we hold the water pistol ($\mathbf{p}_1$), it is impossible to aim it in an admissible manner to hit the target (to allow voters to get a better outcome).

In the Gibbard–Satterthwaite Theorem, we know the "aiming" directions – in a level set where c_1 is chosen, a voter without c_1 top-ranked wants to move to a level set choosing a preferred c_j. Now that we know the direction of profile changes, we want to choose level sets (in profile space) so that changing profile can't enter the agent's personally more favored level set. By using this geometry to define the level sets, the procedure is defined. But, as true with shooting a water pistol, the target will be wet if the angle between the aiming direction and a normal vector is less than 90°. Indeed, it turns out (Sect. 4.3, Saari 1984) that to avoid manipulation, the level set must be orthogonal to this profile change.

This construction is trivial to accomplish with $n = 2$ alternatives. If the outcome is c_1, then a voter with c_1 top-ranked doesn't want to move. A voter with c_1 bottom-ranked, however, tries to improve c_2's chance. With two alternatives, the only possible change is to vote for c_1 rather than c_2. To make this strategic action counterproductive, choose the level sets so such a move helps c_1 – this is monotonicity! (So, "monotonicity" determines the orientation of the level sets. This ensures that a strategic action moves the wrong way through the "paper towel.") With $n \geq 3$ alternatives, however, $(n-1)/n$ of all $n!$ voter types want to change the outcome; this introduces far too many directions of profile changes to be countered. After all, a typical boundary separating two open sets of profiles (leading to different choices) is one dimension lower than the total space, so at each point there is a unique orthogonal direction. To be strategy proof, the surface must be orthogonal to

[18] For level set approaches, see the paper by Chichilnisky and Heal and by Rasmussen in this collection of papers.

each of these directions; the overabundance of voter types trying to change the outcome require too many orthogonal directions for this to be geometrically possible. In fact, the argument shows that the closest one can come to partially satisfying these conditions is with the Copeland method. (For details, see Saari 1994b.)

From this perspective, the Gibbard–Satterthwaite Theorem belongs to those results where changes in profiles lead to surprising changes in outcome. This includes those paradoxes where, after a candidate receives more support, she does poorer. As these conclusions involve how a profile change passes from one level set into another, they depend on the geometric orientation of the level sets. (A detailed discussion is in [Saari) (1994b, Sect. 4.1.4.2].)

4. Symmetry level sets

Finally, to illustrate how the standard assumption of "neutrality" defines a level set geometry, I selected positional methods.[19] But, this approach applies to many other classes of procedures.

The principal voting symmetry is neutrality; if each voter changes the names of the candidates in a systematic manner, then the original conclusion is similarly permuted. This symmetry action defines the level sets – each level set identifies a particular procedure. For intuition, consider points on the unit sphere. Rotation about the z axis is a symmetry action that spins a point around the axis creating a circle – the *orbit* of the rotation action. When the sphere represents the Earth, these orbits are the latitude lines. An orbit, then, partitions the space into sets that share the common property of the specified symmetry. (For latitude, this "sameness" is the angle formed by points with the z axis). Observe the special role played by the North and South Poles; each defines a *singular orbit* – a lower-dimensional point rather than a line. Also important is the smooth transition to the singular setting; the orbits (the latitude lines) of points near the z axis are small circles.

Neutrality requires a name change for the candidates to be accompanied by a similar change in outcome. Mathematically, if σ is a permutation representing a name change, then $\sigma(\mathbf{p})$ represents the change in the profile obtained when each voter varies the names in the indicated manner. Similarly $\sigma(f(\mathbf{p}))$ changes the names in the outcome (the choice or the ranking) according to the permutation σ. A procedure is neutral if

$$f(\sigma(\mathbf{p})) = \sigma(f(\mathbf{p})). \tag{4.1}$$

To apply neutrality to positional voting with $n \geq 3$ candidates, let the jth axis of R^n represent the tally for $c_j, j = 1, \ldots, n$. Thus, when $\mathbf{w}^n = (w_1, w_2, \ldots, 0)$ is used to tally the ballots, the *vector ballot* for a voter with the preferences $\mathscr{A}^n = c_1 \succ c_2 \succ \ldots \succ c_n$ is the point $\mathbf{w}^n \in R^n$. A voter with different preferences defines a permutation of $\mathscr{A}^n$, $\sigma(\mathscr{A}^n)$. The vector ballot for voter $\sigma(\mathscr{A}^n)$ is

[19] Some of these results are reported in a series of papers, e.g., Saari (1989, 1990, 1992a), but how these results were discovered and an outline of related unpublished conclusions has not; this is done here.

a permutation of $\mathbf{w}^n$ represented by $\mathbf{w}^n_\sigma$. To illustrate with $\mathbf{w}^3 = (2, 1, 0)$, the votes cast for a voter with preferences $\mathscr{A}^3 = c_1 \succ c_2 \succ c_3$ form the vector (2, 1, 0) reflecting that two points are totaled for c_1, one point for c_2 and none for c_3. Similarly, if σ is the permutation of $\mathscr{A}^3$ defining the ranking $c_2 \succ c_3 \succ c_1$ then $\mathbf{w}^3_\sigma = (0, 2, 1)$.

All possible permutations of the n agents defines the group S_n; the orbit of this group, $\{\mathbf{w}^b_\sigma\}_{\sigma \in S_n}$, defines all possible vector ballots. The convex hull of this orbit is

$$\mathscr{C}\mathscr{H}(\mathbf{w}^n) = \left\{ \sum_{\sigma \in S_n} t_\sigma \mathbf{w}^n_\sigma \mid t_\sigma \geq 0, \sum_{\sigma \in S_n} t_\sigma = 1 \right\}. \tag{4.2}$$

If t_σ is interpreted as the portion of all voters with the ranking $\sigma(\mathscr{A}^n)$, then the sum in Eq. (4.2) represents the election outcome. The ranking of the candidates is determined by ranking the values of the coordinates of the sum.

To identify this discussion with the one about points on a sphere, recognize that a scalar multiple of $\mathbf{w}^n$ makes no difference; e.g., the election ranking is the same if (2, 1, 0) or if (200, 100, 0) is used to tally the ballots. Therefore, assume that each voting vector has unit length. In other words, a voting vector defines a direction, or equivalently, a point on a sphere. The choice of (2, 1, 0) becomes, therefore, $5^{-\frac{1}{2}}(2, 1, 0)$. The orbits of the plurality (1, 0, 0) and anti-plurality $(\frac{1}{\sqrt{2}}, \frac{1}{\sqrt{2}}, 0)$ are in Fig. 8a; that of the Borda Count is in Fig. 8b. The shaded region denotes convex hull, or the space of all possible normalized election outcomes – these are the "neutrality level sets" for the procedures.

A major concern in choice theory is to determine how the same voters sincerely rank different subsets of candidates. (This is a justification for IIA.) With $n \geq 3$ candidates, list the $2^n - (n+1)$ subsets with two or more candidates in some order as $D_1, D_2, \ldots, D_{2^n-(n+1)}$ where $|D_j|$ is the number of candidates in D_j. The space of election outcome becomes the product space $R^{|D_1|} \times R^{|D_2|} \times \cdots \times R^{|D_{2^n-(n+1)}|}$ where the labels of the coordinate axis in

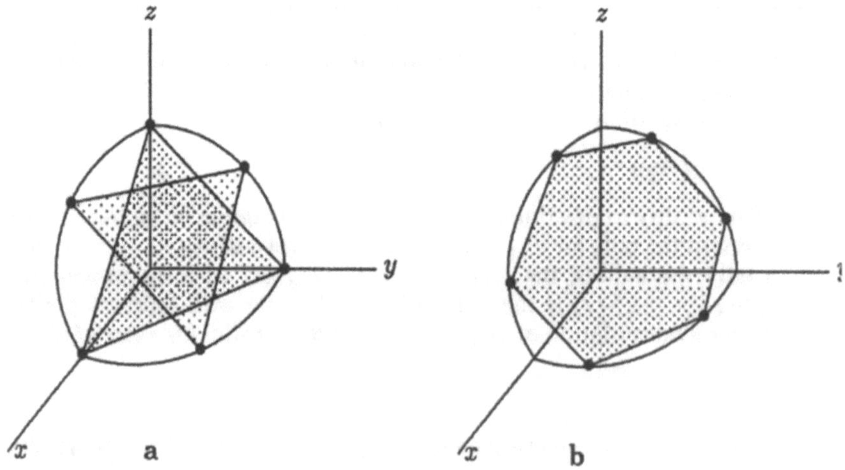

Fig. 8. "Neutrality" level sets

$R^{|D_j|}$ indicate the names of candidates in D_j. For instance, the coordinate x_j^k represents the fate of c_j within subset D_k. In this manner with $n = 3$, the nine-dimensional space has coordinates $(x_1^1, x_2^1; x_1^2, x_3^2; x_2^3, x_3^3; x_1^4, x_2^4, x_3^4)$.

Next, assign to each subset of candidates a normalized voting vector $\mathbf{w}^{|D_j|}$. This collection of voting vectors defines the *system voting vector*

$$\mathbf{W}^n = (\mathbf{w}^{|D_1|}, \mathbf{w}^{|D_2|}, \ldots, \mathbf{w}^{|D_{2^n-(n+1)}|}).$$

The *system vector ballot* for a $\mathscr{A}^n$ voter, $\mathbf{W}^n$, indicates the points this voter assigns to each candidate for each of the $2^n - (n+1)$ subsets of candidates. As above, the *system vector ballot* for a $\sigma(\mathscr{A}^n)$ voter is represented by $\mathbf{W}_\sigma^n$. Again, as above, the election outcome over all possible subsets of elections is a point in the convex hull defined by the orbit $\{\mathbf{W}_\sigma\}_{\sigma \in S_n}$; it is

$$\mathscr{CH}(\mathbf{W}^n) = \left\{ \sum_{\sigma \in S_n} t_\sigma \mathbf{W}_\sigma^n \mid \sum_{\sigma \in S_n} t_\sigma = 1, \ t_\sigma \geqslant 0 \right\}.$$

Thus, the differences in positional voting procedures as well as all properties and consequences of using $\mathbf{W}^n$ are reflected by differences in the algebraic orbit structure of $\mathbf{W}^n$ and the geometry of $\mathscr{CH}(\mathbf{W}^n)$! Again, the basic properties distinguishing procedures are obtained by differences in the level set geometry.

Think of each $\mathbf{W}^n$ as a point on a higher dimensional sphere. This suggests that the orbits defined by the neutrality symmetry action inherit some of the flavor of the construction of latitude lines for the sphere. Indeed, geometric differences in these "neutrality level sets" distinguish the different properties among different system voting vectors. As a simple illustration, the larger the dimension of a set in $R^{|D_1|} \times R^{|D_2|} \times \cdots \times R^{|D_{2^n-(n+1)}|}$, the more kinds of points with previously unused coordinates are admitted. As each coordinate corresponds to different kinds of election outcomes over the subsets of candidates, a larger dimensional subset $\mathscr{CH}(\mathbf{W}^n)$ represents settings where the election outcomes can be quite varied over the different subsets of candidates – these define the kinds and types of admissible paradoxes of voting. Thus, just the dimensional aspects of $\mathscr{CH}(\mathbf{W}^n)$ identify profound consequences about the procedures.[20]

4.1. *The kinds of results*

To see the kinds of results, remember that the dimensionality of $\mathscr{CH}(\mathbf{W}^n)$ has implications about the kinds of properties and paradoxes admitted by $\mathbf{W}^n$. From the "latitude line" description on the sphere, we must identify lower dimension objects with singular orbits. Therefore, it is reasonable to wonder whether some $\mathbf{W}^n$ defines a singular orbit (for the wreath product group action) much like the North and South Poles defines singular settings for the

[20] For the reader familiar with group theory, note that $\{\mathbf{W}_\sigma^n\}_{\sigma \in S_n}$ is not the orbit of the permutation group S_n; it is the orbit of the more interesting *wreath product of permutation groups*. What adds interest is that this group structure should be used to analyze any system which admits neutrality!

rotation action on spheres. There is; *it occurs when the Borda count (BC) is assigned to each subset of candidates.*

To indicate the cascade of new conclusions resulting from this singular behavior assertion, notice that since the BC defines a singular orbit, its lower dimension convex hull forces fewer kinds of BC election results to emerge over the different subsets of candidates. But, new kinds of election outcomes correspond to "election paradoxes", so we must expect the BC to admit fewer paradoxes, both in kinds and numbers, than any other $\mathbf{W}^n$. This is the case. (Saari 1990)

To see what else happens, recall from singularity theory that "singularities" of a group action form a "stratification." For intuition, notice that the rank structure of

$$A = \begin{pmatrix} x & 0 \\ 0 & y \end{pmatrix}$$

has A with full rank for all (x, y) in $\Sigma^0 = \{(x, y) | xy \neq 0\}$, rank one in $\Sigma^1 = \{(x, y) \neq \mathbf{0} \,|\, xy = 0\}$, and rank zero in $\Sigma^2 = \{\mathbf{0}\}$. Because there are locations where A has zero and full rank, we know there are locations where A has rank one. Moreover, observe the intimate relations of these singular settings where Σ^2 is in the closure of Σ^1, and Σ^1 is in the closure of Σ^0.

A similar situation holds for the singularity structure of positional voting procedures as created by the BC. Singularity theory tells us that other positional methods exist which fall somewhere between the BC and the "worse case" scenario for voting (which includes the plurality vote). This ensures the existence of a lower dimensional set of system voting vectors that admit different levels of paradoxes and election properties for the outcomes over the different subsets of candidates. This set of voting vectors, α^n, has the structure of an algebraic set. (Saari 1989)

From the containment structure of the closures of the Σ^j's in the matrix example, we must expect the system voting procedures to have a structure where sets with lower dimensional orbits are in the closure of sets with larger dimensional orbits. And, while there are discontinuities in the properties admitted by different procedures, one must expect the kinds of paradoxes admitted by procedures to be related as one moves to higher dimensional sets. In fact, by exploiting the stratified singularity structure of these orbits, it becomes possible to define a partial ordering over system voting vectors to identify which procedures admit more kinds of paradoxes, how they are related, and why. (Saari 1992a, 1993)

4.2. The likelihood of paradoxes

A standard topic from choice theory is to compare procedures in terms of how likely it is that they admit desired or undesirable outcomes. These answers are surprisingly easy to extract from the orbit structures. After all, to measure the likelihood of a certain outcome, say that a Condorcet winner is not elected, we want to describe that portion of the convex hull $\mathcal{CH}(\mathbf{W}^n)$ allowing this behavior. But, we now have geometric means to provide much sharper answers. As an indication of this, observe that the latitude circles on the sphere shrink in size as the base point approaches the singular positions (the North and South Poles). Similarly, the size of the convex hull $\mathcal{CH}(\mathbf{W}^n)$ continuously shrinks to that of the BC hull as each component of $\mathbf{W}^n$ approaches the BC voting vector. Consequently, the closer a procedure approximates the BC, the closer (in probabilistic and other terms) it is to achieving the desired BC properties.

4.3. New kinds of paradoxes

As a final indication of consequences from this geometric approach, consider the common sense notion that, in general, lower dimensional objects are invariant with respect to additional operations and symmetries. Consequently, we must (accurately) expect the BC to satisfy more kinds of symmetry relationships than any other positional voting procedure. For instance, neutrality requires that when all voters change the names of the voters in the same manner, the outcome changes according to this same permutation. Similarly, suppose after all voters mark ballots by listing the candidates from top to bottom, they discovered they were wrong – the ranking should have to be reversed with higher ranked candidates listed towards the bottom. As all voters completely reversed their ranking of the candidates, it is reasonable to expect that the election outcome is similarly reversed. However, for $n = 3$, *only the BC respects this reversal symmetry*! Indeed, for any other procedure, it can be that the election tally for a profile and its reversal remain unchanged! (A similar property holds for all $n \geq 3$.) The extra dimensions of the other hulls, which prohibits this symmetry, is manifested by a new class of paradoxes! (See Saari 1994b for $n = 3$ examples and an explanation.)[21]

5. Summary

A way to unravel the complexity of choice issues is to analyze the geometry of the associated information. Even with primitive geometric tools (e.g., crude aspects of the orientation of surfaces), answers for important questions are forthcoming. With increased sophistication (e.g., the orbits of positional methods), sharper results become available. So, while geometric approach toward information is in an early developmental stage, already it is providing answers and insight for a host of questions.

[21] Mathematically, neutrality defines a subgroup (of order $n!$) of the permutation group $S_{n!}$; it identifies all ways there are to permute the $n!$ voter types. The reversal operation, however, is *not* in this subgroup. On the other hand, the BC commutes with the larger subgroup of $S_{n!}$ generated by neutrality and the reversal operation. Indeed, the full *Borda subgroup*, the subgroup of $S_{n!}$ that the BC commutes with, is even larger for $n > 3$.

References

Arrow KJ (1963) Social choice of individual values, 2nd edn. Wiley, New York
Baigent NJ, Gaertner W (1993) Never choose the uniquely largest: a characterization. preprint.
Chichilnisky G (1982) The topological equivalence of the Pareto condition and the existence of a dictator. J Math Econ 30: 223–233
Chichilnisky G, Heal G (1983) Necessary and sufficient conditions for a resolution of the social choice paradox. J Econ Theory 31: 68–87
Chichilnisky G, Heal G (1997) The geometry of implementation: a necessary and sufficient condition for straightforwardness. Soc Choice Welfare 14: 259–294
Fishburn P (1970) Arrow's impossibility theorem: Concise proof and infinite voters. J Econ Theory 2: 103–106
Gibbard AF (1973) Manipulation of voting schemes; a general result. Econometrica 41: 587–601
Haunsperger D (1992) Dictionaries of paradoxes for statistical tests on k samples. J Am Stat Assoc 87: 149–155
Hansson B (1976) The existence of group preferences. Public Choice 28: 89–98
Rasmussen H (1997) Strategy-proofness of continuous aggregation maps. Soc Choice Welfare 14: 249–257
Saari DG (1984) A method for constructing message systems for smooth performance functions. J Econ Theory 33: 249–274
Saari DG (1987) Chaos and the theory of elections. In: Kumshanski A, Sigmunol K (eds) Dynamical Systems: Lecture notes in Economics and Mathematical Systems, vol. 287, Springer, pp 179–188
Saari DG (1988) Symmetry, voting and social choice. Math Intelligencer 10: 32–42
Saari DG (1989) A dictionary for voting paradoxes. Econ Theory 48: 443–475
Saari DG (1990) The Borda Dictionary. Soc Choice Welfare 7: 279–317
Saari DG (1991a) Erratic behavior in economic models. J Econ Behav Organ 16: 3–35
Saari DG (1991b) Calculus and extensions of Arrow's Theorem. J Math Econ 20: 271–306
Saari DG (1992a) Symmetry extensions of "neutrality" I: advantage to the Condorcet loser. Soc Choice Welfare 9: 307–336
Saari DG (1992b) Millions of election rankings from a single profile. Soc Choice Welfare 9: 277–306
Saari DG (1993) Symmetry extensions of "neutrality" II: Partial ordering of dictionaries. Soc Choice Welfare 10: 301–3
Saari DG (1994a) Inner consistency or not inner consistency; a reformulation is the answer. In: Barnett W, Moulin H, Salles M, Schfield N (eds) Social Choice and Welfare
Saari DG (1994b) Geometry of voting. Springer-Verlag, Berlin Heidelberg, New York
Satterthwaite MA (1975) Strategyproofness and Arrow's conditions: existence and correspondence theorems for voting procedures and social welfare functions. J Econ Theory 10: 187–217
Sen A (1977) Social choice theory: a re-examination. Econometric 45: 53–89
Young P (1974) An axiomatization of Borda's Theory. J Econ Theory 9: 478–486

Topological methods in social choice: an overview

Paras Mehta

Balliol College, Oxford OX1 3BJ, United Kingdom

Received: 30 December 1994/Accepted: 22 April 1996

Abstract. This paper gives an exposition of the topological framework for social choice theory developed by Chichilnisky, and reviews the mathematical concepts needed for understanding this framework. Within that context, this paper also discusses some classic results of Chichilnisky.

I. Introduction

In Arrow's framework for social choice theory, a society of k individuals seeks to decide among a set X of alternatives, where X is simply a listing of the discrete alternatives with no additional mathematical structure. Each individual, then, has a preference over X, consisting of a complete, reflexive, and transitive binary relation over X, and a social choice function gives a rule for aggregating any particular profile of individual preferences into a statement of group preference.

In this paper, however, we adopt Chichilnisky's topological framework for social choice, which is motivated by the concept of economic decision making taking place over a commodity space such as Euclidean $\mathbb{R}^n$, rather than a discrete set of alternatives. In this framework, the basic social choice scenario is formulated as follows:

We have a society of $k \geq 2$ individuals who are faced with deciding among a set of alternatives in the choice space (or commodity space) $X = \mathbb{R}^n$, $n \geq 2$. Associated with each individual is a preference p over X, where p is a codimension one, oriented foliation of X, so that the leaves of p define the indifference hypersurfaces of the preference, and the orientation indicates direction of increasing preference. Alternatively, we can view p as a normalized (to unit Euclidean length) vector field over X, where for each $x \in X$, $p(x)$ is the unit vector normal to the leaf of the foliation passing through x, and pointing in the direction given by the orientation of the foliation. At each $x \in X$, then, $p(x)$ is the vector normal to the indifference hypersurface of the preference, and indicating the direction of increasing preference in the choice space. We will further require that the preferences be continuously differentiable and locally integrable, so that the space of all possible preferences P is simply the space of all continuously differentiable, locally integrable unit-normalized vector fields over X.

A social choice function, then, is a map $f: P^k \to P$ that associates to each k-tuple of preferences $\mathbf{p} = (p_1, \ldots, p_k)$, representing individually the k members of society, an aggregate preference $f(\mathbf{p})$ representative of the entire society.

In this paper, we begin by discussing some of the tools of algebraic topology, and homotopy theory in particular, used in topological social choice theory. Then, we review some classical results of Chichilnisky investigating the possibility of requiring social choice functions to satisfy various conditions of ethics and rationality.

II. Results from algebraic topology

To each topological space Y (which we will assume for simplicity is path connected) we can associate homotopy groups $\pi_i(Y)$, where the i-dimensional homotopy group $\pi_i(Y)$ "counts" the number of i-dimensional "spherical holes" in Y. Consider, for instance, the first homotopy group $\pi_1(Y)$. We fix a point y_0 in Y and look at paths in Y which begin and terminate at y_0. Then, we consider the equivalence classes of paths under the relation which says that for two paths γ_1 and γ_2, $\gamma_1 \sim \gamma_2$ if γ_1 can be continuously deformed into γ_2. Formally, a path γ is a continuous map $\gamma: [0, 1] \to Y$ with $\gamma(0) = \gamma(1) = y_0$. Then $\gamma_1 \sim \gamma_2$ if there exists a continuous map $G: [0, 1] \times [0, 1] \to Y$ with $G(x, 0) = \gamma_1(x)$ and $G(x, 1) = \gamma_2(x)$. The set of paths in Y modulo this equivalence relation defines $\pi_1(Y)$, with the group operation being composition of paths; i.e., for paths γ_1, γ_2, $\gamma_1 \circ \gamma_2$ will be that path obtained by traversing γ_1 first and then γ_2.

For example, in $\mathbb{R}^2$, any path can be continuously deformed into the constant path which maps $[0, 1]$ into the single point y_0. Hence, all paths are equivalent, and $\pi_1(\mathbb{R}^2)$ is trivial. Now puncture the plane at the origin, so that $Y = \mathbb{R}^2 \backslash \{0\}$. If we consider a path γ which traverses a circle around the origin once in a clockwise direction, we can see that this path cannot possibly be deformed into the constant path because we cannot collapse the circle into a single point without passing through the origin. Similarly, a circular path that passes around the origin twice counterclockwise cannot be continuously deformed into a path passing around the origin only once, nor can it be continuously deformed into a path passing around the origin twice clockwise. In fact, a path in Y can be identified, up to the homotopy equivalence relation, by the number of times it passes around the origin, and the direction in which it does so. Hence, $\pi_1(Y)$ is simply the infinite cyclic group of integers $\mathbb{Z}$ under addition. Note that π_1 here has one generator, and Y has precisely one 1-dimensional hole.

For our purposes, it will be useful to use a different definition of paths: a path γ can be seen as a continuous map $\gamma: S^1 \to Y$, where S^1 is the one-dimensional unit sphere (i.e., the unit circle). Using this definition and the appropriate homotopy equivalence relation ($\gamma_1 \sim \gamma_2$ if there exists a continuous deformation $G: S^1 \times [0, 1] \to Y$ with $G(x, 0) = \gamma_1(x)$ and $G(x, 1) = \gamma_2(x)$), we obtain the same homotopy group $\pi_1(Y)$. Then, we define the higher homotopy groups in an analogous fashion: $\pi_i(Y)$ consists of equivalence classes of paths $\gamma: S^i \to Y$ where two paths are equivalent if they can be continuously deformed into one another. And, in analogous fashion, these will measure the number of higher dimensional holes in Y.

In this paper we need only concern ourselves with the homotopy groups of i-dimensional spheres S^i and of Cartesian products of j i-dimensional spheres $(S^i)^j$. We will use the fact that $\pi_i(S^i) = \mathbb{Z}$. This makes some intuitive sense: S^i has one i-dimensional hole, so we would expect $\pi_i(S^i)$ to be the group generated on one generator. Furthermore, from a more general result which states that the homotopy group of a Cartesian product of path connected spaces is the direct sum of the homotopy groups of the individual factor spaces, we have that $\pi_i((S^i)^j) = \oplus_{k=1}^{j} \pi_i(S^i) = \oplus_{k=1}^{j} \mathbb{Z}$. Essentially, this says that elements of the homotopy group of a product of spheres can be identified uniquely by the action of the coordinate projection maps.

Now if we have a continuous map $m: X \to Y$, this induces a map $m^*: \pi_i(X) \to \pi_i(Y)$ on the homotopy groups which maps $\alpha \in \pi_i(X)$ to $\beta \in \pi_i(Y)$ as follows: If $f: S^i \to X$ is a path contained in the class of paths α, then β is the class of paths containing $m \circ f$. In fact, the unique map m^* is a homomorphism; i.e., it commutes with the composition of group elements.

As we noted above, $\pi_i(S^i) = \mathbb{Z}$ (under addition), so any map $m: S^i \to S^i$ between spheres defines a homomorphism of the integers $m^*: \mathbb{Z} \to \mathbb{Z}$. Under these circumstances, we say that $m^*(1)$ is the degree of the map m, which we shall denote $d(m)$. Intuitively, the degree tells us how many times the sphere in the domain is "wrapped around" the sphere in the range. For example, consider a map $m: S^1 \to S^1$. If m is the identity map, its degree is 1 since the domain sphere is wrapped around the range sphere precisely once. Now we think of S^1 as the unit circle in $\mathbb{R}^2$ and let m be the map that takes the vector forming angle θ with the positive horizontal axis to the vector forming angle $2\pi - \theta$. In this case, the domain sphere is wrapped once around the range sphere, but the orientation is reversed; i.e., as we traverse the domain sphere counterclockwise, we traverse the range sphere clockwise. Hence, the degree of this map is -1, negative because it is orientation reversing. Another way of thinking about the degree, given the above notion of wrapping the domain sphere around the range sphere, is that the degree counts the number of points in the preimage of a "suitably nice" point in the range, taking into account whether the map preserves or reverses orientation.

There are a few facts we will want to invoke about the degree of maps. First, given maps $m_1, m_2: S^i \to S^i$, if m_1 is homotopic to m_2 (i.e., if m_1 can be continuously deformed into m_2 via a homotopy $G: S^1 \times [0, 1] \to S^1$ with $G(x, 0) = m_1(x)$, $G(x, 1) = m_2(x)$), then the degree of m_1 and m_2 are equal. That is to say, the degree is invariant among maps in the same homotopy class, and in fact, the degree defines a homomorphism of $\pi_i(S^i)$ into $\mathbb{Z}$. Second, we note that any nonsurjective continuous map on spheres is of degree zero.

III. An induced map on spheres

We now return to our social choice function $f: P^k \to P$, for aggregating the preferences of $k \geq 2$ individuals over the choice space $X = \mathbb{R}^n$, $n \geq 2$. In this section, we construct a map $\phi:(S^{n-1})^k \to S^{n-1}$ induced by the social choice function $f: P^k \to P$ and examine the degree of ϕ when restricted to subspaces homeomorphic to a sphere. The following construction and argument is due to Chichilnisky (e.g., in [3], where it is used to prove the non-existence of continuous, majority-type rules).

For any $v \in S^{n-1}$, we denote by λ_v the linear preference given by the direction v; i.e., λ_v is the constant vector field with $\lambda_v(x) = v$ for all $x \in X$ (alternatively, the linear preference λ_v is the foliation whose leaves are the hyperplanes with normal v and whose orientation is given by the direction of v). Then, we define the map $\lambda: S^{n-1} \to P$ which continuously embeds S^{n-1} in P by mapping $v \in S^{n-1}$ to the linear preference λ_v. Finally, we fix a point $y \in X$ and define $\phi:(S^{n-1})^k \to S^{n-1}$ by

$$\phi:(v_1, \ldots, v_k) \mapsto [f(\lambda_{v_1}, \ldots, \lambda_{v_k})](y).$$

One can interpret ϕ as acting, in accordance with f, on the linear preferences given by $v_1, \ldots, v_k$, and then approximating the preference $f(\lambda_{v_1}, \ldots, \lambda_{v_k})$ by a linear preference at the point y. Alternatively, one can simply view ϕ itself as a social choice function on the subspace of linear preferences. In any case, ϕ is continuous by the continuity of λ, of f, and of preferences $f(\lambda_{v_1}, \ldots, \lambda_{v_k})$ over X.

Next, fix a point $v_0 \in S^{n-1}$ and consider the following subspaces of $(S^{n-1})^k$. Denote by D the "diagonal" of $(S^{n-1})^k$ given by $D = \{(v, v, \ldots, v, v) : v \in S^{n-1}\}$, and by E_i the "ith edge" of $(S^{n-1})^k$ given by $E_i = \{(v_1, \ldots, v_i, \ldots, v_k) : v_j = v_0$ for $j \neq i$, and $v_i \in S^{n-1}\}$ (in the case $n = 2, k = 2$, $(S^{n-1})^k = S^1 \times S^1$ is a torus and can be depicted as a square with edges identified, and then D and E_1, E_2 are the diagonal and edges of the square, respectively). Finally, we define the following canonical inclusion and projection maps:

$$\tau_D: S^{n-1} \to (S^{n-1})^k, \qquad \tau_D: v \mapsto (v, \ldots, v) \in D,$$

$$\tau_i: S^{n-1} \to (S^{n-1})^k, \qquad \tau_i: v \mapsto (v_0, \ldots, v_0, v_i, v_0, \ldots, v_0) \in E_i,\ v_i = v,$$

$$\varrho_i: (S^{n-1})^k \to S^{n-1}, \qquad \varrho_i: (v_1, \ldots, v_k) \mapsto v_i.$$

These maps induce the homomorphisms τ_D^*, τ_i^*, and ϱ_i^*, respectively, on $(n-1)$th homotopy groups. The identity map on any domain will be denoted ι.

Our next step is to investigate the degree of ϕ restricted to the subspaces D, E_i by examining the homotopy classes of ϕ restricted to those subspaces and invoking the homotopy invariance of the degree discussed in the last section. We make some preliminary observations prior to examining the degree of $\phi|_D$ and of $\phi|_{E_i}$, denoted $d(\phi|_D)$ and $d(\phi|_{E_i})$. First, D and each E_i are homeomorphic to S^{n-1}, so it is reasonable to speak of the degree of these maps. Strictly speaking, however, references to $d(\phi|_D)$ and $d(\phi|_{E_i})$ actually mean $d(\phi \circ \tau_D)$ and $d(\phi \circ \tau_i)$, respectively, but we will feel free to use the two notations interchangeably. Finally, the remaining discussion in this section is based solely on the topology of $(S^{n-1})^k$, with no particular properties of f or ϕ (except, of course, continuity) invoked.

Maps enclosed in brackets ([ ]) denote members of $(n-1)$th homotopy groups; since the maps we will consider have domain S^{n-1}, by our comments in Sect. II, we can consider these maps as representative of homotopy group elements. It will be clear from context whether the homotopy classes are elements of $\pi_{n-1}(S^{n-1})$ or $\pi_{n-1}((S^{n-1})^k)$.

Recalling that the maps τ_i are the canonical inclusions of S^{n-1} into $E_i \subset (S^{n-1})^k$, we next choose a map $m: S^{n-1} \to (S^{n-1})^k$ such that $[m] = \sum_{j=1}^{k} [\tau_j] \in \pi_{n-1}((S^{n-1})^k)$. Then, for all i, $[\varrho_i \circ m] = \varrho_i^*([m])$, by the definition of the induced homomorphism, and

$$\varrho_i^*([m]) = \varrho_i^*\left(\sum_{j=1}^{k} [\tau_j]\right) = \sum_{j=1}^{k} \varrho_i^*([\tau_j]) = [\varrho_i \circ \tau_i] \in \pi_{n-1}(S^{n-1}). \tag{1}$$

The penultimate equality follows from the fact that ϱ_i^* is a homomorphism. The last equality follows from the fact that $\varrho_i([\tau_j]) = [\varrho_i \circ \tau_j]$, and $\varrho_i \circ \tau_j$ is a constant mapping onto v_0 for $i \neq j$, so that $[\tau_j]$ is in the kernel of ϱ_i^* for $i \neq j$. Moreover, since $\varrho_i \circ \tau_D \equiv \varrho_i \circ \tau_i \equiv \iota: S^{n-1} \to S^{n-1}$ (the identity map) for all i, we conclude that $[\varrho_i \circ \tau_D] = [\varrho_i \circ \tau_i] = [\iota]$ for all i, and hence by Eq. (1) that for all i

$$[\varrho_i \circ m] = [\varrho_i \circ \tau_i] = [\varrho_i \circ \tau_D] = [\iota] \in \pi_{n-1}(S^{n-1}).$$

But since $[\varrho_i \circ m] = [\varrho_i \circ \tau_D]$ means precisely that $\varrho_i^*([m]) = \varrho_i^*([\tau_D])$ (for all i), we conclude that $[m] = [\tau_D]$, i.e., that m is homotopic to τ_D. This follows as a consequence of the fact cited in Sect. II that elements of $\pi_{n-1}((S^{n-1})^k)$ are uniquely determined by the action of the projection homomorphisms ϱ_i^*. The composition of homotopic maps with homotopic maps produces composites which are themselves homotopic, and hence

$$[\phi \circ m] = [\phi \circ \tau_D]. \tag{2}$$

Recalling that $[m] = \sum_{j=1}^{k} [\tau_j]$, we have that

$$[\phi \circ m] = \phi^*([m]) = \phi^*\left(\sum_{j=1}^{k} [\tau_j]\right) = \sum_{j=1}^{k} \phi^*([\tau_j]) = \sum_{j=1}^{k} [\phi \circ \tau_j]. \tag{3}$$

From the homotopy invariance of degree, and the fact that the degree defines a homomorphism between $\pi_{n-1}(S^{n-1})$ and $\mathbb{Z}$, Eqs. (2) and (3) give us the desired conclusion:

$$\sum_{j=1}^{k} d(\phi \circ \tau_j) = d(\phi \circ \tau_D), \text{ or equivalently,} \tag{4}$$

$$\sum_{j=1}^{k} d(\phi|_{E_j}) = d(\phi|_D). \tag{5}$$

This line of argument, culminating in Eqs. (4) and (5), is not altogether suprising intuitively if we note that the diagonal D can be continuously deformed into the union of the edges $\bigcup_{j=1}^{k} E_j$. For example, consider the simplest case, when $n = 2, k = 2$. Then we have D, E_1, $E_2 \subset S^1 \times S^1$. But $S^1 \times S^1$ is just the torus, which we can consider as a square in the plane with sides identified; in this case, D is the diagonal of the square, and E_1, E_2 perpendicular edges. Then we clearly see that by "stretching" the diagonal to one pair of perpendicular edges, we can continuously deform D into $E_1 \cup E_2$. This gives some insight into why the degree of ϕ restricted to the diagonal should be equal to the sum of the degrees of ϕ restricted to each edge.

IV. Selected topological results in social choice theory

This section is dedicated to presenting and proving three classic results, due to Chichilnisky, concerning the possibility of requiring a social choice function to satisfy certain selected criteria of ethics and rationality. In all three results, we will require that a social choice function $f: P^k \to P$ be continuous. We can view continuity as a requirement of stability on the social choice functions, so that a suitably small change in individual preferences does not produce a wild change in the aggregate preference of the society.

A social choice function f is said to satisfy *unanimity* if and only if f satisfies an "identity" condition along the "diagonal" of P^k; formally, unanimity requires that for all preferences $p \in P$, $f(p, p, \ldots, p, p) = p$. Unanimity merely ensures that the aggregation rule defined by f has some minimal relationship to the preferences on which it acts. An *anonymous* social choice function is invariant under permutation of the k entries in its argument; i.e.,

$$f(p_1, \ldots, p_k) = f(p_{\pi(1)}, \ldots, p_{\pi(k)})$$

for all permutations π and all profiles of preferences $(p_1, \ldots, p_k)$. Anonymity offers a strong condition of egalitarianism in which the rule of preference aggregation takes note only of the preferences that are held by individuals in the society, and not of the identity of the particular person who holds a specific preference. The following impossibility result is due to Chichilnisky [1, 2, 4]:

Theorem 1. *No continuous social choice function $f: P^k \to P (k \geq 2)$ satisfies the conditions of unanimity and anonymity.*

Proof. Suppose that a continuous social choice function f respecting unanimity and anonymity exists. Then consider the continuous mapping $\phi: (S^{n-1})^k \to S^{n-1}$ induced by f; ϕ inherits respect of unanimity and anonymity from f:

$$\phi(v, \ldots, v) = f(\lambda_v, \ldots, \lambda_v)(y) = \lambda_v(y) = v$$

and

$$\begin{aligned}\phi(v_{\pi(1)}, \ldots, v_{\pi(k)}) &= f(\lambda_{v_{\pi(1)}}, \ldots, \lambda_{v_{\pi(k)}})(y) \\ &= f(\lambda_{v_1}, \ldots, \lambda_{v_k})(y) = \phi(v_1, \ldots, v_k)\end{aligned}$$

for all permutations π.

By Eq. (5), we have that

$$\sum_{j=1}^{k} d(\phi|_{E_j}) = d(\phi|_D).$$

Now $d(\phi|_D) = 1$, since by unanimity $\phi \circ \tau_D$ is the identity, and the degree of the identity map is 1. And by anonymity, the degree on each edge is the same: $d(\phi|_{E_i}) = d(\phi|_{E_j}) = \lambda$ for some $\lambda \in \mathbb{Z}$ and for all i, j. But now we have that $k\lambda = 1$ for some $\lambda \in \mathbb{Z}$, which has no solutions in λ over $\mathbb{Z}$ for integral $k \geq 2$. □

One way of overcoming the impossibility result of Theorem 1 is to relax the conditions placed on f. For example, we can certainly construct a continuous social choice function $f: P^k \to P$ that respects unanimity; simply pick a dictatorial rule. Say, for instance, that $f:(p_1, \dots, p_k) \mapsto p_1$. Then this rule, which makes the first individual a dictator, satisfies both continuity and unanimity. We make one further observation about this social choice function. If we consider the induced map $\phi:(S^{n-1})^k \to S^{n-1}$ (which is also continuous and respects unanimity) at any point, then ϕ restricted to E_1 and D both "cover" S^{n-1} once, suggesting that both have degree one, while ϕ restricted to E_j, $j \neq 1$, is simply a constant map (and hence of degree zero). This is precisely the result we would expect given our study of the degree of ϕ (in particular, Eq. (5)).

Producing a continuous, anonymous social choice function is a bit trickier. For ease of description, we restrict our attention to the case $k = 2$ and $n = 2$. Furthermore, we consider only the space of linear preferences $P_L \cong S^1$. Then a social choice function is a mapping on the torus $f: S^1 \times S^1 \to S^1$. If we consider S^1 to be the unit circle centered about the origin in $\mathbb{R}^2$, then an element of S^1 is given by an angle, and we can define f by angular addition. Then f is clearly anonymous and continuous, and as we would expect, f restricted to each edge covers the image S^1 once, and f restricted to the diagonal covers the image twice.

Rather than weakening either anonymity or unanimity, another way of overcoming the impossibility result of Theorem 1 is to restrict the domain P of possible preferences. For example, Chichilnisky and Heal [5] show that a necessary and sufficient condition for the existence of continuous, anonymous, and unanimous social choice functions $f: P^k \to P$, for all $k \geq 2$, is that the space of preferences P be topologically contractible. In [6], Heal considers several examples of contractible preference spaces and interprets contractibility in terms of some sort of "limited agreement." For example, if everyone always agrees on some direction in the commodity space which is undesirable (if there exists some direction $v \in S^{n-1}$ such that for all preferences $p \in P$, $p(x) \neq v$ for all x in the commodity space), then P is contractible.

In the next result, we consider continuous social choice functions that satisfy a stronger requirement than that of unanimity called the *Pareto criterion*. Of course, in light of Theorem 1, such a social choice function is definitely not anonymous. The Pareto criterion on $f: P^k \to P$ simply requires that, for any k-tuple of preferences $(p_1, \dots, p_k) \in P^k$, if $x \in X$ is preferred to $y \in X$ according to each p_i, $1 \leq i \leq k$, then x is preferred to y according to $f(p_1, \dots, p_k)$. This definition naturally demands some explanation of what it means for one alternative x to be preferred to y according to a preference p. Since we have assumed our preferences to be locally integrable, we can view p as the gradient, locally, of a utility function, and define comparability, locally, in terms of that function. Specifically, suppose at each $x \in X$ there exist a neighborhood N of x and a utility function $u: N \to \mathbb{R}$ such that $p|_N \equiv \nabla u$. Then for any $y \in N$, we say $x \geq_p y$ if and only if $u(x) \geq u(y)$.

We now note the following consequence of the Pareto criterion (as noted in [3, 4]): At any point $x \in X$, if $\mathbf{p} = (p_1, \dots, p_k) \in P^k$ satisfies the condition that for some $v, w \in S^{n-1}$, $p_i(x) = v$ or $p_i(x) = w$ for all i, then $f(\mathbf{p})(x)$ is contained in the "cone" spanned by v and w. Specifically, suppose that we have a $\mathbf{p}$ such that, at $x \in X$, $p_i(x) = v$ or w for some $v, w \in S^{n-1}$ and for all i. We think of S^{n-1} as an object centered at the origin in Euclidean $\mathbb{R}^n$. Then, $p_i(x)$ for each i is the normal to the surface formed by those points $y \in X$ such that preference p_i is indifferent between choices x and y. Such an indifference surface separates (in some local neighborhood) X into two disjoint subsets of those choices $y \in X$ that are strictly preferred to x by p_i, and of those choices $y \in X$ to which x is strictly preferred by p_i. Hence, for all $\varepsilon > 0$ sufficiently small, this means that

$$x + \varepsilon \cdot v \geq_{p_i} x$$

for all $v \in S^{n-1}$ satisfying $\langle v, p_i(x) \rangle \geq 0$, where $\langle \cdot, \cdot \rangle$ is the standard Euclidean inner product. Now we return to the case under consideration, in which each $p_i = v$ or w at $x \in X$. The preceding discussion implies that for all $\varepsilon > 0$ sufficiently small, $x + \varepsilon \cdot v$ is preferred to x by all p_i for all $v \in S^{n-1}$ having nonnegative inner product with both v and w. By the Pareto criterion, then, $f(\mathbf{p})(x)$ must have nonnegative inner product with all $v \in S^{n-1}$ having nonnegative inner product with v and w. In particular, this means precisely that $f(\mathbf{p})(x)$ is contained in the closed cone generated by v and w. Moreover, this consequence implies respect of unanimity (just take $v = w$ in the preceding argument).

We further note that this consequence, that $f(\mathbf{p})(x)$ lies in the cone generated by v, w if $p_i(x) = v$ or w for all i, is strictly weaker than the Pareto criterion itself. In fact, it is the only consequence of the Pareto criterion that we will use in this section and the next. Therefore, in Theorems 2 and 3 below, we can replace the requirement that f satisfy the Pareto criterion with the requirement that f satisfy the general consequence of the Pareto criterion discussed above.

Finally, we define the *nonnegative association axiom*. Suppose that there exist a point $x \in X$, a k-tuple of preferences $\mathbf{p} = (p_1, \dots, p_k)$, and an index i, such that $f(\mathbf{p})(x) = -p_i(x)$. Then f satisfies the nonnegative association axiom if and only if for any k-tuple of preferences $\mathbf{q} = (q_1, \dots, q_k)$ such that $q_j(x) = -q_i(x)$ for all $j \neq i$, $f(\mathbf{q})(x) \neq q_i(x)$. Intuitively, the nonnegative association axiom states that whenever everyone in the society has the same preference locally at some alternative $x \in X$, with the exception of a single individual who has the exact antipodal preference, the aggregate preference should not locally at x coincide with the single individual in the minority *if* there is even one k-tuple of preferences which gives rise to an aggregate that, at x, is antipodal to that same single individual in the minority of one. The axiom can be seen as a very mild condition of rationality or consistency. Then we have the following characterization of Chichilnisky [4].

Theorem 2. *Let $f: P^k \to P$ be a continuous social choice function satisfying the Pareto criterion and the nonnegative association axiom. Then f is homotopic to a dictatorial social choice function.*

As its name suggests, a dictatorial social choice function is simply one that projects $\mathbf{p} = (p_1, \dots, p_k)$ onto its ith coordinate for a fixed i.

Proof. The main strategy behind this proof is to prove that under the hypotheses of the theorem, there is some coordinate i such that $f(\mathbf{p})$ is never equal to $-p_i$, and then to conclude that f is homotopic to the map $\tilde{f}$ which projects the profile $(p_1, \dots, p_k)$ onto p_i.

We fix a point $y \in X$, a direction $v_0 \in S^{n-1}$, and consider the induced map ϕ: $(S^{n-1})^k \to S^{n-1}$ and the subspaces D and E_i. By Eq. (5), $\sum_{j=1}^{k} d(\phi|_{E_j}) = d(\phi|_D)$. By our previous discussion, respect of the Pareto criterion implies that f, and hence ϕ, respects unanimity. So, $\phi \circ \tau_D$ is the identity on S^{n-1}, and $d(\phi|_D) = 1$.

Now we consider $d(\phi|_{E_i})$. The point $\phi(v_0, \ldots, v_0, v_i, v_0, \ldots, v_0)$ must be contained in the cone spanned by v_0 and v_i by the Pareto criterion. But this, together with continuity, implies that $\phi(v_0, \ldots, v_0, -v_0, v_0, \ldots, v_0) \in \{v_0, -v_0\}$. *A priori*, according to Pareto, $\phi(v_0, \ldots, v_0, -v_0, \ldots, v_0)$ can be any $v \in S^{n-1}$, since the cone generated by $v_0, -v_0$ consists of all of S^{n-1}. However, suppose that $\phi(v_0, \ldots, v_0, -v_0, v_0, \ldots, v_0) = w \notin \{v_0, -v_0\}$. Then consider a sequence $\{v^j\}_{j=1}^{\infty}$ converging to $-v_0$ and such that the closed cone generataed by v_0 and v^j does not contain w for any j (this is possible because the cone generated by two points on the sphere is contained within a hemisphere unless those two points are antipodal). Then $\lim_{j\to\infty} v^j = -v_0$, but $\lim_{j\to\infty} \phi(v_0, \ldots, v_0, v^j, v_0, \ldots, v_0) \neq w$, in violation of continuity. So we conclude that $\phi(v_0, \ldots, v_0, -v_0, v_0, \ldots, v_0) \in \{v_0, -v_0\}$.

Now if $v_i \neq -v_0$, then $\phi(v_0, \ldots, v_0, v_i, v_0, \ldots, v_0) \neq -v_0$, since for $v_i \neq -v_0$, the cone generated by v_0, v_i is contained in $S^{n-1} \backslash \{-v_0\}$. Hence, if $\phi(v_0, \ldots, v_0, -v_0, v_0, \ldots, v_0) = v_0$, then $\phi|_{E_i}^{-1}(-v_0) = \emptyset$, $\phi|_{E_i}$ is not surjective, and hence its degree is zero. On the other hand, if $\phi(v_0, \ldots, v_0, -v_0, v_0, \ldots, v_0) = -v_0$, then $\phi|_{E_i}^{-1}(-v_0) = \{(v_0, \ldots, v_0, -v_0, v_0, \ldots, v_0)\}$ and $d(\phi|_{E_i}) = 1$.

By Eq. (5), and since $d(\phi_D) = 1$, there exists an index i^* such tht

$$d(\phi|_{E_i}) = \begin{cases} 1 & \text{if and only if } i = i^*, \\ 0 & \text{if and only if } i \neq i^*, \end{cases} \tag{6}$$

or, equivalently,

$$\phi|_{E_i}^{-1}(-v_0) = \begin{cases} (v_0, \ldots, v_0, v_{i^*}, v_0, \ldots, v_0),\ v_{i^*} = -v_0 & \text{if and only if } i = i^*, \\ \emptyset & \text{if and only if } i \neq i^*. \end{cases} \tag{7}$$

Moreover, this implies that

$$\phi_{\cup_{i=1}^k E_i}(v_1, \ldots, v_k) \neq -v_{i^*}. \tag{8}$$

To understand why (8) must be the case, consider first the edge E_i when $i \neq i^*$. Then for all $(v_1, \ldots, v_k) \in E_i, -v_{i^*} = -v_0$; but by (7), for $i \neq i^*$, $\phi|_{E_i}(E_i) \cap \{-v_0\} = \emptyset$, or in other words, $-v_{i^*} \notin \phi|_{E_i}(E_i)$ when $i \neq i^*$. Next we consider the edge E_i when $i = i^*$. In this case v_{i^*} is allowed to vary over S^{n-1}, so we have two possibilities. First, when $v_{i^*} \neq -v_0$, then $\phi|_{E_{i^*}}(v_0, \ldots, v_{i^*}, \ldots, v_0) \neq -v_{i^*}$ since $-v_{i^*}$ is not in the cone spanned by v_0 and v_{i^*} as long as $v_{i^*} \neq -v_0$. Second, when $v_{i^*} = -v_0$, then by Eq. (7), $\phi|_{E_i^*}(v_0, \ldots, v_0, v_{i^*}, v_0, \ldots, v_0) = v_{i^*}$. So we conclude (8).

But now we have, following the above reasoning, that for any $v_0' \in S^{n-1}$, if we define $E_i' = \{(v_1, \ldots, v_k) \in (S^{n-1})^k \colon v_j = v_0' \text{ for } j \neq i, \text{ and } v_i \in S^{n-1}\}$, there is an index j^* satisfying

$$\phi_{\cup_{i=1}^k E_i'}(v_1, \ldots, v_k) \neq -v_{j^*}, \tag{9}$$

analogous to Eq. (8) above. Moreover, by continuity considerations, we must have that $j^* = i^*$, for if $j^* \neq i^*$, we would have

$$\phi(v_0, \ldots, v_0, \overbrace{-v_0,}^{=v_{i^*}} v_0, \ldots, v_0) = -v_0,$$

$$\phi(v_0', \ldots, v_0', \overbrace{-v_0',}^{=v_{i^*}} v_0', \ldots, v_0') = v_0',$$

in violation of continuity (the first equality follows from Eq. (7); the second follows from continuity, Pareto, and Eq. (9)).

Furthermore, given any $\mathbf{v} = (v_1, \ldots, v_k) \in (S^{n-1})^k$, if $\phi(\mathbf{v}) = -v_{i^*}$, then by the nonnegative association axiom, $\phi(-v_{i^*}, \ldots, -v_{i^*}, v_{i^*}, -v_{i^*}, \ldots, -v_{i^*}) = -v_{i^*}$, which contradicts the property of i^* established in (8) and (9). Hence, we must have that $\phi(\mathbf{v}) \neq -v_{i^*}$ for all $\mathbf{v} \in (S^{n-1})^k$.

So far we have been considering a map ϕ induced at a point $y \in X$. The above discussion implies that for any $y' \in X$, there exists a map ϕ' analogous to ϕ, and an index i' analogous to i^*. Specifically, we have that $\phi'(v_1, \ldots, v_k) \neq -v_{i'}$. Again, however, continuity considerations require that $i' = i^*$, for if not, for some $w \in S^{n-1}$:

$$\phi(w, \ldots, w, \overbrace{-w,}^{=v_{i^*}} w, \ldots, w) = -w,$$

$$\phi'(w, \ldots, w, \overbrace{-w,}^{=v_{i^*}} w, \ldots, w) = w,$$

or alternatively, recalling the construction of ϕ, ϕ' and the use of λ to denote linear preferences,

$$f(\lambda_w, \ldots, \lambda_w, -\lambda_w, \lambda_w, \ldots, \lambda_w)(y) = -w,$$

$$f(\lambda_w, \ldots, \lambda_w, -\lambda_w, \lambda_w, \ldots, \lambda_w)(y') = w,$$

violating the continuity of the preference that is the image of f.

All of this leads to the conclusion that there exists an index i^* such that $f(p_1, \ldots, p_k)(y) \neq -p_{i^*}(y)$ for all k-tuples of preferences $\mathbf{p} \in P^k$ and all $y \in X$. In other words, if we denote by $\tilde{f}$ the dictatorial rule $\tilde{f}(p_1, \ldots, p_k) = p_{i^*}$, then we have $f(\mathbf{p}) \neq -\tilde{f}(\mathbf{p})$ for all $\mathbf{p} \in P^k$. But now we have that f is homotopic to $\tilde{f}$ by the homotopy $F: P^k \times [0, 1] \to P$:

$$F(\mathbf{p}, t)(y) = \frac{(1-t)f(\mathbf{p})(y) + t\tilde{f}(\mathbf{p})(y)}{\|(1-t)f(\mathbf{p})(y) + t\tilde{f}(\mathbf{p})(y)\|}. \qquad \square$$

In Theorem 3, we present the final characterization of social choice functions in this paper, again using the construction of Sect. III. The social choice functions under consideration in Theorem 3 are continuous and satisfy the Pareto criterion. We also introduce a new condition called the *decisive majority axiom.* Essentially, the decisive majority axiom requires that, whenever a k-tuple of preferences $\mathbf{p} = (p_1, \ldots, p_k)$ can be partitioned locally at a point $x \in X$ into two classes that hold completely opposite preferences from one another, then the aggregate $f(\mathbf{p})$ locally at x will agree with the majority. Formally, for all $x \in X$ and for all $\mathbf{p} = (p_1, \ldots, p_k) \in P^k$, if there exist a number r and a direction $v \in S^{n-1}$ such that $\#\{p_i : p_i(x) = v\} = r$ and $\#\{p_i : p_i(x) = -v\} = k - r$, and if $r > k - r$, then $f(\mathbf{p})(x) = v$.

We make some observations about the decisive majority axiom. First, it is strictly weaker than a majority rule requirement. Second, we will not, in fact, use the full power of the decisive majority axiom. We will only be interested in the axiom for the case $r = k - 1$; in fact, we can replace the decisive majority axiom in the statement of Theorem 3 with the requirement that the decisive majority axiom need hold only in the case $r = k - 1$. We then have the following [3].

Theorem 3. *No continuous social choice function satisfies the Pareto criterion and the decisive majority axiom.*

In particular, since the decisive majority axiom is implied by a majority rule requirement, Theorem 3 immediately implies that there is no continuous, Pareto, majority rule social choice function. This consequence should not be surprising: in a majority rule decision-making process, a single change in preferences can lead to a large change in outcome, at odds with the demands of continuity.

Proof. Suppose $f : P^k \to P$ is continuous and satisfies the Pareto criterion and decisive majority axiom. Then fix some $v_0 \in S^{n-1}$, $y \in X$, and consider the induced map ϕ and subspaces D, E_i. By Eq. (5), $d(\phi|_D) = \sum_{j=1}^{k} d(\phi|_{E_j})$. Since the Pareto criterion implies unanimity, $d(\phi|_D) = 1$.

Now consider $d(\phi|_{E_i})$. By Pareto, $\phi|_{E_i}(v_0, \ldots, v_0, v_i, v_0, \ldots, v_0) = w$ must be in the cone generated by v_0, v_i. In particular, $w \neq -v_0$ except possibly when $v_i = -v_0$. But $\phi|_{E_i}(v_0, \ldots, v_0, -v_0, v_0, \ldots, v_0) = -v_0$ would violate the decisive majority axiom. So we conclude that $\text{image}(\phi|_{E_i}) \subseteq S^{n-1} \backslash \{-v_0\}$. Then $\phi|_{E_i}$ is not surjective, and hence $d(\phi|_{E_i}) = 0$ for all i. The last statement gives the desired contradiction:

$$1 = d(\phi|_D) = \sum_{j=1}^{k} d(\phi|_{E_i}) = 0. \qquad \square$$

Acknowledgements. The author wishes to thank Daniel Goroff, Graciela Chichilnisky, Geoffrey Heal, and Amartya Sen for their assistance in this study of social choice theory.

References

1. Chichilnisky G (1979) On fixed point theorems and social choice paradoxes. Econ Lett 3: 347–351
2. Chichilnisky G (1980) Social choice and the topology of the spacse of preferences. Adv Math 37: 165–176
3. Chichilnisky G (1982) Structural instability of decisive majority rules. J Math Econ 9: 207–221
4. Chichilnisky G (1983) Social choice and game theory: recent results with a topological approach. In: Pattanaik PK, Salles M (eds) Social Choice and Welfare. North-Holland, Amsterdam
5. Chichilnisky G, Heal G (1983) Necessary and sufficient conditions for a resolution of the social choice paradox. J Econ Theory 31: 68–87
6. Heal G (1983) Contractibility and public decision-making. In: Pattanaik PK, Salles M (eds) Social Choice and Welfare. North-Holland, Amsterdam

A note on continuous social choice

Yuqing Zhou

Graduate School of Business, Columbia University, New York, NY 10027, USA

Received: 28 February 1994/Accepted: 22 April 1996

Abstract. We discuss continuous social choice functions defined on the set of the equivalence classes induced by profiles of individual preferences corresponding to anonymity. It is shown that we can at best obtain the equivalent result of Chichilnisky's impossibility theorem when the set of an equivalence classes is endowed with various reasonable topologies.

1. Introduction

In her pioneering papers, Chichilnisky (1979, 1980, 1982) develops a topological framework within which Arrow's impossibility theorem is extended. There are several comments and extensions following her work: Chichilnsky and Heal (1983) provides necessary and sufficient conditions under which a social choice function satisfying Chichilnisky rule exists; Baigent (1984, 1985) tries to reinterpret Chichilnisky's impossibility theorem in the quotient space of profiles of individual preferences induced by an alternative approach. Other extensions include Baigent (1987, 1989), Baigent ahd Huang (1990), Chichilnisky (1993), Chichilnisky and Heal (1984), LeBeton and Uriarte (1990), etc. The reader may refer to them for details.

In this note we make several comments on Baigent's alternative approach. At first glance, it seems quite appealing to restrict the domain of social choice functions to the set of the equivalence classes, since any function defined on this domain automatically satisfies anonymity. As such, we might obtain a very nice function space to work with. However, the definition of continuity in this new framework become problematic, i.e., there is no intuitive way to define a topology on this alternative domain. It is shown that, based on

The author is grateful to G. Chichilnisky and G. Heal for helpful comments and suggestions. The research was done while the author was studying at Stanford University.

Baigent's alternative approach, we can at best expect an equivalent result of Chichilnisky's impossibility theorem, which is the case when the set of equivalence classes is endowed with the quotient topology. In general, when the set of equivalence classes is endowed with other reasonable topologies, we can only expect weaker results in a sense which will be made clear shortly.

2. Chichilnisky's framework

Let X be a topological choice space and let $I = \{1, 2, \ldots, n\}$ denote a set of individuals. A preference p is a binary relation on X which is connected, reflexive and transitive. P will denote the set of all preferences. A profile $\bar{p}$ is an n-tuple of preferences; that is $\bar{p} = (p_1, p_2, \ldots, p_I)$. Pn denotes the set of all profiles; that is $Pn = P \times P \times \times \ldots \times P$, n times. A social choice function is a function $f: Pn \to P$.

We may endow the space of preferences P with various topologies (see Chichilnisky 1979, 1980). Given any topology σ_p on P, a continuous social choice function may be defined with respect to the product topology on Pn induced by σ_p. A social choice function f is said to respect unanimity if $f(p, p, \ldots, p) = p$, $p \in P$. It is called anonymous if f is invariant with respect to any permutation of the values of its arguments. The well-known Chichilnisky's impossibility theorem may be stated as follows.

Theorem 2.1 (Chichilnisky's impossibility theorem) *There does not exist a continuous social choice function which is anonymous and respect unanimity when P is not contractible.*

Baigent (1984, 1985) questions the properness of continuity based on the product topology. He claims that Chichilnisky's formulation is unsatisfactory in view of the justification for anonymity. His argument goes as follows: Assume that σ_p is induced by a metric δ on P. Then the corresponding product topology on Pn is induced by a metric d on Pn such that

$$d(\bar{p}, \bar{q}) = \sum_{i=1}^{n} \delta(p_i, q_i)$$

He argues that d does give a significant role to the assignment of preferences to individuals because $d(\bar{p}, \bar{q}) > 0$ if $\bar{p} \neq \bar{q}$ even if both $\bar{p}$ and $\bar{q}$ are permutations of each other, and that the justification of anonymity requires that a zero distance is assigned to any two profiles constituted from the same set of preferences. Therefore, he proposes an alternative approach in which the continuity is defined on the quotient space of profiles of individual preferences induced by the equivalence class corresponding to anonymity, and shows that Chichilnisky's impossibility theorem still persists.

3. An equivalent result

It is easy to see that the above-mentioned result is equivalent to Chichilnisky impossibility theorem. To make it precise, let

$$\Omega = \{f \in C(Pn, P) \mid f(p, p, \ldots, p) = p, f(p_1, p_2, \ldots, p_n) = f(p_{\beta(1)}, \ldots, p_{\beta(n)})\},$$

where β is any permutation of I and $C(Pn, P)$ represents the set of all continuous functions between Pn (with product topology) and P. Let

$$\bar{\Omega}_q = \{\bar{f} \in C([Pn], P) \mid \bar{f}([p, p, \ldots, p]) = p\},$$

where $[Pn]$ denote the set of equivalence classes induced by Pn corresponding to anonymity, and $[Pn]$ is endowed with quotient topology. Let $\pi : Pn \to [Pn]$ be the canonical map. The equivalent theorem can be stated as follows:

Theorem 3.1. *There exists a one-to-one correspondence between Ω and $\bar{\Omega}_q$, which is defined by $\Phi : \Omega \to \bar{\Omega}_q$, where $\Phi(f) = f \circ \pi^{-1}$ and the inverse map Φ^{-1} is defined by $\Phi^{-1}(\bar{f}) = \bar{f} \circ \pi$. Thus Ω is empty if and only if $\bar{\Omega}_q$ is empty.*

Sketch of proof: It is easy to see that $\Phi^{-1}(\bar{f}) \in \Omega$ since both π and $\bar{f}$ are continuous functions. As for $\Phi(f)$, we note that for any open set U in P

$$f^{-1}(U) = \pi^{-1}[\pi f^{-1}(U)] = \pi^{-1}[\Phi(f)^{-1}(U)]$$

is open in Pn. Furthermore, $[Pn]$ is endowed with the quotient topology, which implies that $\Phi(f)^{-1}(U)$ is open in $[Pn]$. Thus $\Phi(f) \in \bar{\Omega}_q$. ■

Note that in the proof of Theorem 3.1 we used the fact that $[Pn]$ is endowed with the quotient topology, which is the finest topology that makes the canonical map π continuous. It seems that we can also endow $[Pn]$ with other topologies. Certainly it is desirable for us to focus on those topologies which make the canonical map continuous, and those topologies will be weaker than the quotient topology. Therefore, one may only expect weaker results in these cases. Let $\bar{\Omega}_w$ be the same as $\bar{\Omega}_q$ except that the quotient topology is replaced by a weaker topology.

Theorem 3.2. *There exists an injection map Ψ from $\bar{\Omega}_w$ to Ω which is defined by $\Psi(\bar{f}) = \bar{f} \circ \pi$. Thus Ω is empty implies that $\bar{\Omega}_w$ is empty. But the inverse is not true*

Proof. (Sketch) The proof is similar to that of Theorem 3.1. ■

4. Concluding remarks

Baigent's argument is misleading since anonymity is characterized by the function f: $Pn \to P$, not by the product topology for profiles. Thus his seemingly different approach is at best equivalent to Chichilnisky's approach. Naturally no new insight is provided in his alternative approach. And in most cases, the results are not so impressive, as theorem 3.2 has shown.

In sum, it is not a good idea to work on the alternative framework. Chichilnisky's original framework contains more information and has more economic intuition than the alternative one.

References

Baigent N (1984) A reformulation of Chichilnisky's impossibility theorem. Econ Lett 16: 23–25

Baigent N (1985) Anonymity and continuous social choice. J Math Econ 13: 1–4
Baigent N (1987) Preference proximity and anonymous social choice. Q J Econ 102: 162–169
Baigent N (1989) Some further remarks on preference proximity. Q J Econ 104: 191–193
Baigent N, Huang P (1990) Topological social choice. Soc Choice Welfare 7: 141–146
Chichilnisky G (1979) On fixed point theorems and social choice paradoxes. Econ Lett 3: 347–351
Chichilnisky G (1980) Social choice and the topology of spaces of preferences. Adv Math 37: 165–176
Chichilnisky G (1982) Social aggregation rules and continuity. Q J Econ 96: 337–352
Chichilnisky G, Heal G (1983) Necessary and sufficient conditions for a resolution of the social choice paradox. J Econ Theory 31: 68–87
Chichilnisky G, Heal G (1984) Patterns of power. J Public Econ XV: 177–184
LeBreton M, Uriarte JR (1990) On the robustness of the impossibility result in the topological approach to social choice. Soc Choice Welfare 7: 131–140

Strategy-proofness of continuous aggregation maps*

Heine Rasmussen

Norwegian School of Economics and Business Administration, Helleveien 30, N-5035 Bergen-Sandviken, Norway

Received: 28 February 1994/Accepted: April 22, 1996

Abstract. We consider continuous aggregation maps (e.g., social welfare functions). By assuming that the voters have preferences over social outcomes, we regard the social decision procedure as a noncooperative game, with the aggregation map as a game form. The map is called strategy-proof if it is a Nash equilibrium that the voters report their most preferred outcomes. We give sufficient topological conditions on the space of outcomes so that only dictatorial maps are strategy-proof.

1. Introduction

In this paper we consider strategy-proofness of continuous social welfare functions and other aggregation procedures. Similar problems have been examined by several authors, surveyed by Pattanaik [6] and Sen [7]. The results we present here are closely related to two articles by Chichilnisky [1] and Chichilnisky and Heal [3]. Their work on this and related issues differs from most of the literature by the requirement that the social welfare function shall be continuous.

This, of course, presupposes that a topology is defined on the space of preferences. If we (for the moment) confine the study to linear preferences over $\mathbf{R}^n$ (i.e., preferences that can be represented by nonzero linear utility functions), the orthogonal vectors to the indifference surfaces will have the same direction everywhere. The space of linear preferences can then be given the same topology as S^{n-1} (the $n-1$ dimensional sphere) by identifying any preference with the point where the orthogonal vector at the origin of $\mathbf{R}^n$ intersects an

* The author is grateful to Graciela Chichilnisky and Geoffrey Heal for helpful discussions, and to Terje Lensberg for comments on the manuscript.

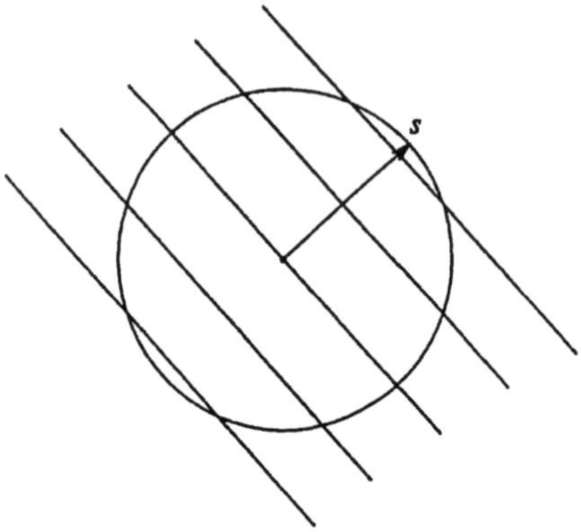

Fig. 1. This construction maps a linear preference to a point s in S^1

$n-1$ dimensional sphere centered at the origin of $\mathbf{R}^n$. The two-dimensional case is illustrated in Fig. 1.

With linear preferences over $\mathbf{R}^n$ thus regarded as points in S^{n-1}, a continuous social welfare function defined for m voters becomes a map from the product of m copies of S^{n-1} to the space of social preferences, which we assume is identical to the space of individual preferences, i.e., S^{n-1}.

For instance, if we assume that the social welfare function is defined for two voters with preferences over $\mathbf{R}^2$, it can be regarded as a map from the torus to the circle. Call this map $f: S^1 \times S^1 \to S^1$. It can be shown that the degree of the restriction of f to the diagonal of $S^1 \times S^1$ is equal to the sum of the degrees of the restrictions of f to the two subspaces $x \times S^1$ and $S^1 \times x$, where x is an arbitrary point in S^1. If the restriction of f to the diagonal is of nonzero degree (e.g., if f respects unanimity), then at least one of the last two restrictions of f must be nonzero degree. From the fact that any map of nonzero degree is onto, it is easily seen that at least one of the two agents can always achieve any social outcome he may desire, given the preference the other agent reports (in general, of course, he will have to misrepresent his preferences to accomplish this). In the terminology of Chichilnisky and Heal [3], such an agent is called a strategic dictator.

This argument is easily generalized to spheres of arbitrary dimension, and an arbitrary number of voters, and the result then becomes:

Theorem 1 (Chichilnisky and Heal). *If a continuous social welfare function* $f: (S^n)^m \to S^n$ *(where* $n \geq 1$ *and* $m \geq 2$*) is of nonzero degree when restricted to the diagonal, there is always a strategic dictator.*

The proof, using degree arguments, depends on the spherical structure of the preference space. However, one can easily imagine spaces of preferences that are not homeomorphic to an n-sphere. In several papers, e.g. Chichilnisky and Heal [2] and Heal [4], it has been shown that *noncontractibility* of the preference space gives rise to social choice paradoxes. It is thus natural to ask whether there is a connection between these paradoxes and the result above; in other words, can Theorem 1 (or some version of it) be extended to noncontractible spaces in general? The rest of this paper is devoted to that question.

2. Two counterexamples

So far, we have been referring to spaces of preferences. However, since the mathematical results presented here can be applied to other problems than aggregation of preferences (for some examples, see Chichilnisky and Heal [3]), we will henceforth consider just an abstract topological space X. We furthermore assume that there is defined an *aggregation map* $f: X \times X \to X$, to be interpreted in the following way: Two agents each report a point in X (their "votes"), and the aggregation map then selects a social outcome in X. In order to keep notation and arguments as simple as possible, we will not consider cases with more than two agents, even though there are some generalizations to an arbitrary number of agents.

Let both agents have preferences[1] over the points of X. It will be assumed that these preferences have unique global maxima. If we regard the agents' preferences over X as preferences over social outcomes, the aggregation map gives rise to a two-person noncooperative game, where the possible moves for both players are the points in X, and with outcome $f(x_1, x_2)$ if the players' moves are x_1 and x_2, respectively. We say that the aggregation map is *strategy-proof* for a given pair of preferences over X if it is a Nash equilibrium in this game that both players report their most preferred point (the unique maximum of their preferences).

In general, an aggregation map can clearly be strategy-proof for some pairs of preferences, but not for other pairs. By a *maxima-exhaustive class of preferences over* X we mean a collection of preferences with unique maxima, such that every point in X is the maximum of some preference in the collection. Given a maxima-exhaustive class of preferences, we say that f is strategy-proof for this class if it is strategy-proof for any pair of preferences where both preferences are members of the class. In the following, when we say that the map is strategy-proof without referring to a particular class, we mean that the map is strategy-proof for at least *some* maxima-exhaustive class of preferences.

We say that an aggregation map is *dictatorial* (with agent m as a dictator) if

$$f(x_1, \ldots, x_m, \ldots, x_n) = x_m \quad \text{for all } x_1, \ldots, x_n.$$

It is an easy consequence of Theorem 1 that given an arbitrary maxima-exhaustive class of preferences over the points in S^n, the only aggregation maps $f: S^n \times S^n \to S^n$ that may be strategy-proof for this class are the dictatorial ones (assuming the map is of nonzero degree when restricted to the diagonal).

---

[1] Note that the preferences we consider here have no connection with the preferences we described in Section 1. There it was assumed that the result of the social decision process (the social outcome) was a social preference; the preferences we introduce here are individuals' preferences over social outcomes, whether these outcomes are social preferences, or something else. If X is regarded as a space of social preferences, the preferences we describe in this section should then be interpreted as preferences over social preferences.

There is one requirement in the formulation of Theorem 1 that cannot immediately be generalized to other spaces than S^n, and that is the nonzero degree condition. We will replace this with a stronger criterion; we require that the aggregation map shall respect unanimity, i.e., $f(x, x) = x$ for all x. For the spaces S^n, this implies that the restriction of f to the diagonal is of degree 1. Theorem 1 then implies:

Theorem 1′. *Let X be homeomorphic to S^n for some $n \geq 1$. For an arbitrary maxima-exhaustive class of preferences over X, if an aggregation map respects unanimity and is strategy-proof, it is dictatorial.*

Since we use unanimity instead of nonzero degree, the theorem does not refer to any concepts specific to S^n, and we shall try to extend this weaker version of Theorem 1 to a wider class of spaces than the n-dimensional spheres. However, a simple counterexample shows that any attempt to extend it to all noncontractible spaces will fail. Let X be homeomorphic to a cylinder, and define a maxima-exhaustive class of preferences over X as follows: Let d: $X \times X \to \mathbf{R}$ be the natural metric on X, i.e., the distance between two points is defined to be the Euclidean length of the shortest path in X that connects the points. To any point $x \in X$, we associate a preference over X that has x as its unique global maximum, and where $y \gtrsim z$ if $d(x, y) \leq d(x, z)$.

The cylinder can be written as $S^1 \times I$, where I is the unit interval. Denote points in S^1 by σ (with subscripts to distinguish between different points), and points in I by ι. A point $x_1 \in X$ can then be denoted (σ_1, ι_1). Consider the aggregation map f: $X \times X \to X$ defined by

$$f[(\sigma_1, \iota_1), (\sigma_2, \iota_2)] = (\sigma_1, \iota_2).$$

This map is strategy-proof for the maxima-exhaustive class of preferences described above, but it is not dictatorial. An example is depicted in Fig. 2.

We assume that agent 1's most preferred point is x_1, and agent 2's most preferred point is x_2. For each of the two corresponding preferences, we have drawn the indifference curve that contains $f(x_1, x_2)$. We see that agent 1 has noting to gain by reporting anything else than x_1, since he can only achieve outcomes that lie on the horizontal dashed line in the figure (assuming that agent 2 reports x_2). The situation is the same for agent 2, since he can only achieve outcomes on the vertical dashed line.

We have here a noncontractible space, with a very natural class of preferences and a nondictatorial aggregation map, where neither of the agents have incentives to report anything else than their most preferred points.

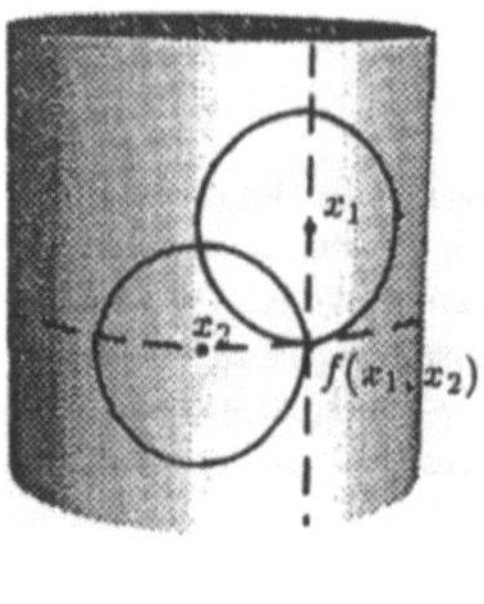

Fig. 2

Consequently, Theorem 1′ does not extend to noncontractible spaces in general. We also observe that the cylinder is homotopy equivalent to the circle, where Theorem 1′ is valid, so the validity of the theorem must depend on topological characteristics below the homotopy level. The obvious question is then: Which of the topological properties that are possessed by the circle, and not by the cylinder, are relevant to our problem?

As a first attempt, we note that the proof of Theorem 1 relies heavily on the important fact that any map from S^n to S^n of nonzero degree is onto. A suitable generalization of this property to topological spaces in general will be to require that any map from X to X that is homotopic to the identity map on X, is onto. We shall call a space X with this property *a retracted space.* However, adding the requirement that X must be a retracted space is still not enough to ensure that only dictatorial maps are strategy-proof for an arbitrary maxima-exhaustive class of preferences. We use the torus as a counterexample.

Let X be homeomorphic to the torus. X can then be written as $S^1 \times S^1$, and points in X will be denoted by (σ, σ'). Let $\hat{d}$ be the natural metric on the circle, i.e., the distance between two points is the Euclidean length of the shortest path in S^1 that connects the points. Define a metric d on the torus by

$$d[(\sigma_1, \sigma'_1), (\sigma_2, \sigma'_2)] = \sqrt{\hat{d}(\sigma_1, \sigma_2)^2 + \hat{d}(\sigma'_1, \sigma'_2)^2}.$$

Define a maxima-exhaustive class of preferences on X by exactly the same method as in our previous counterexample. Let the aggregation map be defined by

$$f[(\sigma_1, \sigma'_1), (\sigma_2, \sigma'_2)] = (\sigma_1, \sigma'_2).$$

The situation is illustrated in Fig. 3, which is to be interpreted in the same way as Fig. 2.

Since the torus is a retracted space, this example shows that even retracted spaces admit strategy-proof aggregation maps that are not dictatorial. However, if we also require that the space shall be an H'-space (to be defined in the next section), it turns out that we get an interesting generalization of Theorem 1′.

3. Sufficient conditions

In the following, all spaces are assumed to be locally contractible. Before we give the definition of an H'-space, we shall explain a few mathematical terms.

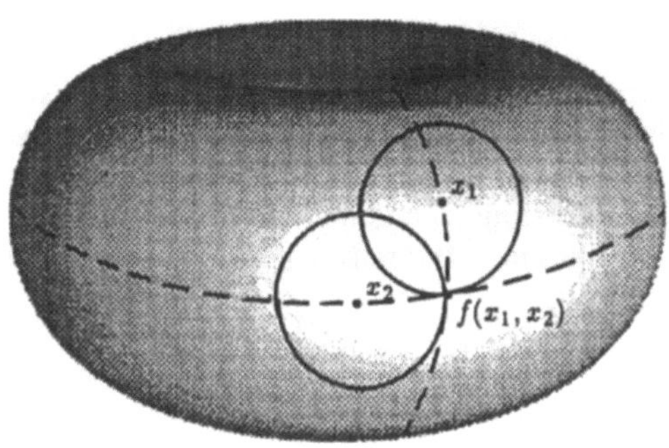

Fig. 3

A *pointed space* is a topological space where an arbitrary point, called the base point, is singled out. The $\vee$*-union* of two pointed spaces X and Y, written $X \vee Y$, is constructed from the disjoint union of the spaces by identifying the two basepoints. Informally, the two spaces can be thought of as being "glued together" at the base points, and at all other points they retain their original topology. More formally, if X has base point x_0 and Y has base point y_0, $X \vee Y$ may be regarded as the subspace $X \times y_0 \cup x_0 \times Y$ of $X \times Y$. E.g, the figure eight can be regarded as the $\vee$-union of two circles.

A *based map* from a pointed space to a pointed space is a map that takes the base point in the first space to the base point in the second. If X, Y, and Z are pointed spaces, two based maps $f: X \to Z$ and $g: Y \to Z$ can be combined into a map $(f, g): X \vee Y \to Z$ by letting f apply to the $X \times y_0$ part of $X \vee Y$, and g to the $x_0 \times Y$ part.

An H'-space[2] is a pointed space X together with a map $\mu: X \to X \vee X$ with the property that the composite maps (where c is the constant map that maps everything to the base point, and i is the identity map)

$$X \xrightarrow{\mu} X \vee X \xrightarrow{(c,i)} X \quad \text{and} \quad X \xrightarrow{\mu} X \vee X \xrightarrow{(i,c)} X \tag{1}$$

are both homotopic to i. Some examples of H'-spaces will be given in Section 4.

If X is an H'-space, we can introduce a binary operator '$*$' on based maps from X to X by defining $f * g$ to be the composite map $(f, g) \circ \mu$, or equivalently, with the diagram

$$X \xrightarrow{\mu} X \vee X \xrightarrow{(f,g)} X.$$

We shall call $f * g$ *the product* of f and g.

Before we state our main result, we shall deduce a general property of strategy-proof aggregation maps. For an aggregation map f and an arbitrary point $x_0 \in X$, define the two maps $f_1^{x_0}, f_2^{x_0}: X \to X$ by

$$f_1^{x_0}(x) = f(x, x_0), \tag{2}$$

$$f_2^{x_0}(x) = f(x_0, x). \tag{3}$$

If we assume that f is strategy-proof, we must have

$$f(x_1, x_2) = f(f(x_1, x_2), x_2) \quad \text{for all } x_1, x_2. \tag{4}$$

This is a consequence of the following argument: Let x_1 and x_2 be two arbitrary points in X, and let agent 1's and agent 2's most preferred points be $f(x_1, x_2)$ and x_2, respectively. Since f by assumption is strategy-proof, we may suppose that agent 2 reports x_2. By reporting x_1, agent 1 can always achieve $f(x_1, x_2)$ (his most preferred point). But since the map is strategy-proof, the

[2] See for instance Maunder [5].

outcome $f(f(x_1, x_2), x_2)$ must be at least as good for him. Considering that $f(x_1, x_2)$ is the unique maximum of his preference, this implies (4). By a symmetric argument we have

$$f(x_1, x_2) = f(x_1, f(x_1, x_2)) \tag{5}$$

It is now an immediate consequence of (4) and (5) that $f_1^{x_0}$ and $f_2^{x_0}$ both are *retractions*; that is, they are equal to the identity map when restricted to their respective images.

We can now prove our main result:

Theorem 2. *Let X be a path-connected, retracted H'-space. For an arbitrary maxima- exhaustive class of preferences over X, if an aggregation map respects unanimity and is strategy-proof, it is dictatorial.*

Proof. Let $f: X \times X \to X$ be an aggregation map that respects unanimity. Our aim is to show that if f is strategy-proof, it must be dictatorial.

Pick an arbitrary base point $x_0 \in X$. Consider the three maps f_Δ, $f_1^{x_0}$, $f_2^{x_0}$: $X \to X$, where f_Δ is defined as

$$f_\Delta(x) = f(x, x),$$

and $f_1^{x_0}$ and $f_2^{x_0}$ are defined by (2) and (3). We first show that f_Δ is homotopic to $f_1^{x_0} * f_2^{x_0}$. Let $u: X \vee X \to X \times X$ be the inclusion map $((i, c), (c, i))$. Call the two composite maps in (1) ϕ and ϕ', respectively. By the definition of the product of two functions, $f_1^{x_0} * f_2^{x_0}$ is equivalent to the composite map

$$X \xrightarrow{\mu} X \vee X \xrightarrow{u} X \times X \xrightarrow{f} X, \tag{6}$$

that is, $f \circ ((i, c), (c, i)) \circ \mu$. An equivalent way of writing this is

$$f \circ ((i, c) \circ \mu, (c, i) \circ \mu),$$

or shorter, $f \circ (\phi, \phi')$. The following diagram will thus define the same composite map as (6):

$$X \xrightarrow{(\phi, \phi')} X \times X \xrightarrow{f} X.$$

This map must clearly be homotopic to f_Δ, since both ϕ and ϕ' are homotopic to i, and if we replace both ϕ and ϕ' in the above diagram with i, the composite is seen to be identical to f_Δ.

It is clear from the definition of the product that for any two maps g, h: $X \to X$, we have $(g * h)(X) \subset g(X) \cup h(X)$. Since $f_1^{x_0} * f_2^{x_0}$ is homotopic to f_Δ (which is equal to the identity map, since f respects unanimity), it must be onto, because X is a retracted space. We thus have

$$f_1^{x_0}(X) \cup f_2^{x_0}(X) = X.$$

We now show by contradiction that one of $f_1^{x_0}$ and $f_2^{x_0}$ must be a constant map that maps everything onto x_0. Assume this is not the case; we can then find a point x_1 different from x_0 so that $x_1 \in f_1^{x_0}(X)$, which implies $f(x_1, x_0) = x_1$. But then we have $x_0 \notin f_2^{x_1}(X)$, as the converse would imply $f(x_1, x_0) = x_0$. By the same argument there is a point x_2 different from x_0 so that $x_0 \notin f_1^{x_2}(X)$. By the properties of X and the fact that $f_1^{x_0} * f_2^{x_0}$ is homotopic to the identity map, we must have

$$f_1^{x_2}(X) \cup f_2^{x_1}(X) = X.$$

We then arrive at a contradiction, since the above argument implies that x_0 is not a member of this union.

This means that one of $f_1^{x_0}$ and $f_2^{x_0}$ is a constant map. Without loss of generality, assume this is true of $f_2^{x_0}$. The map $f_1^{x_0}$ must then be onto, since $f_1^{x_0}(X) \cup f_2^{x_0}(X) = X$. As $f_1^{x_0}$ is a retraction, and is onto, it is the identity map on X. For any x, f_1^x must be homotopic to $f_1^{x_0}$ (since X is path-connected), hence onto (since X is a retracted space), hence f_1^x must be the identity map. But then agent 1 is a dictator. This completes the proof. □

Remark. It is not hard to see that a strategy-proof aggregation map that is onto, must also respect unanimity. If the map is onto, then, for any point x, there must exist points x_1 and x_2 such that $f(x_1, x_2) = x$. This means that $f_1^{x_2}(x_1) = x$, and since $f_1^{x_2}$ is a retraction, we must have $f_1^{x_2}(x) = x$, i.e., $f(x, x_2) = x$. Again this means that $f_2^x(x_2) = x$, and $f_2^x(x) = x$, but this is the same as $f(x, x) = x$. The condition of Theorem 2 that the aggregation map should respect unanimity can thus be replaced with a condition that only requires the map to be onto.

4. Concluding remarks

We have seen that a path-connected space that is both a retracted space and an H'-space does not admit nondictatorial, strategy-proof aggregation maps that respect unanimity. The counterexamples in Section 2 illustrate that neither of the two properties are sufficient by themselves. The space in Fig. 2 is an H'-space, but it is not retracted, as the identity map is homotopic to a map from the cylinder to a circle. The space in Fig. 3 is retracted, but it is not an H'-space.

To get an intuitive feeling for the characteristics of H'-spaces, it is instructive to consider an important class of such spaces, the *suspensions*. The suspension of an arbitrary topological space Y is defined to be the quotient space of $Y \times I$ where $Y \times 0$ is identified to one point and $Y \times 1$ is identified to another point. For example, the suspension of a circle is a cylinder with the two ends collapsed into one point each; in other words, a space homeomorphic to a sphere.

The property of being an H'-space is preserved under homotopy equivalence. This means that the figure eight, for instance, is an H'-space, since it is homotopy equivalent to the suspension of a space consisting of three discrete points.

Informally, retracted H'-spaces can be regarded as generalized spheres, in the sense that they retain two properties possessed by the spherical spaces: (a) They do not contain any proper subspaces that are deformation retracts of themselves (so they are "minimal" representatives of their homotopy type), and (b) loosely speaking, there is always a way to fold such a space "around itself" as illustrated in Fig. 4.

Fig. 4. Folding a circle around itself

References

1. Chichilnisky G (1993) On strategic control. Q J Econ 108: 285–290
2. Chichilnisky G, Heal G (1983) Necessary and sufficient conditions for a resolution of the social choice paradox. J Econ Theory 31: 68–87
3. Chichilnisky G, Heal G (1984) Patterns of power: Bargaining and incentives in two-person games. J Public Econ 23: 333–349
4. Heal G (1983) Contractibility and public decision-making. In: Pattanaik PK, Salles M (eds) Social Choice and Welfare. North-Holland, Amsterdam, pp 103–121
5. Maunder CRF (1970) Algebraic Topology. Van Nostrand Reinhold Company, London
6. Pattanaik PK (1978) Strategy and Group Choice. North-Holland, Amsterdam
7. Sen AK (1981) Social choice theory. In: Arrow KJ, Intriligator MD (eds) Handbook of Mathematical Economics, vol. III. North-Holland, Amsterdam

The geometry of implementation: a necessary and sufficient condition for straightforward games*

G. Chichilnisky, G.M. Heal

Program on Information and Resources, Columbia University,
405 Law Memorial Library, New York, NY 10027, USA

Received: 30 December 1994/Accepted: 22 April 1996

Abstract. We characterize games which induce truthful revelation of the players' preferences, either as dominant strategies (straightforward games) or in Nash equilibria. Strategies are statements of individual preferences on R^n. Outcomes are social preferences. Preferences over outcomes are defined by a distance from a bliss point. We prove that *g is straightforward if and only if g is locally constant or dictatorial* (*LCD*), i.e., coordinate-wise either a constant or a projection map *locally* for almost all strategy profiles. We also establish that: (i) If a game is straightforward and respects unanimity, then the map g must be continuous, (ii) Straightforwardness is a nowhere dense property, (iii) There exist differentiable straightforward games which are non-dictatorial. (iv) If a social choice rule is Nash implementable, then it is straightforward and locally constant or dictatorial.

1. A characterization of straightforward games

In classical forms of resource allocation for public goods,[1] efficiency requires accurate information about people's preferences. However, asking individuals to reveal their preferences can lead to a game in which the truth may or may not be the outcome. When is telling the truth the best strategy? Games in which players' best moves are to say the truth, are called *straightforward*. This

---

* The first versions of these results were completed in 1979, and they were then revised and extended in 1980 and 1981. Versions were circulated as Essex working papers under the titles "Incentives to Reveal Preferences", "Incentive Compatibility and Local Simplicity" and "A Necessary and Sufficient Condition for Straightforwardness". Research support from NSF Grants. SES 79-14050, 92-16028 and 91-10460 and the United Kingdom S.S.R.C. is gratefully acknowledged.

[1] Such as those proposed by Lindahl, Bowen and Samuelson.

paper gives necessary and sufficient conditions for a game to be straightforward.

In the search for straightforward games, certain points are obvious. If a player is a dictator, namely if the outcome is determined solely by her preferences, then she has no incentive to misrepresent those preferences. Likewise, if the outcome is constant, independent of the strategy chosen by the player, then there is no incentive to misrepresent either.

The insight offered in this paper is that these two simple and appealing cases serve as a basis for constructing *all* possible straightforward rules: within a certain family of single peaked preferences defined on the choice space $\mathfrak{R}^n$, *a rule is straightforward if, and only if, it is made up by "piecing together" constant rules and dictatorial rules.* Such rules are called *locally constant or dictatorial* (*LCD*), and they can be very different indeed from dictatorial or constant maps. However, *locally* they behave either like a constant function or like a dictatorial function (a projection) almost everywhere. LCD rules have a remarkably simple geometric structure.

The results presented here were developed between 1979 and 1981[2] and have been circulated widely since them. They are based on an intuitive geometric object: the preimage in strategy space of a given outcome. Our approach is unique in that all of our results are proven by reference to this geometric structure, and are valid for any Euclidean space. This geometric structure has proven to be fruitful elsewhere as well: it was adopted later by Saari [20] and by Rasmussen [19] in this volume, and it is also used in our results on "strategic dictators' in Chichilnisky and Heal [12] and in the results on strategic control in Chichilnisky [8]. We are able to do this because we show (in Theorem 1) that any straightforward game with a convex range (implied for example by respect of unanimity) must be continuous. We can therefore work with continuous maps between Euclidean spaces.

Though simple in concept, locally constant or dictatorial (LCD) maps can be quite complex: several examples are constructed here. LCD rules may satisfy desirable features: they can be continuous and anonymous[3] and also respect unanimity.[4] These are the three axioms proposed by Chichilnisky [6, 7] for characterizing desirable social choice rules.[5]

---

[2] These results were widely circulated and presented at conferences and seminars from 1979 to 1982.

[3] A function $f(x_1, x_2, \ldots, x_n) = y$, $f: \mathfrak{R}^n \to \mathfrak{R}$, is anonymous if $f(x_1, x_2, \ldots, x_n) = f(x_{\pi 1}, x_{\pi 2}, \ldots, x_{\pi n})$ where $(\pi 1, \pi 2, \ldots, \pi n)$ is a permutation of the integers 1 to n. A social choice rule with this property does not discriminate between agents on the basis of their identity.

[4] A social choice rule $f(x_1, x_2, \ldots, x_n) = y$, $f: \mathfrak{R}^n \to \mathfrak{R}$, respects unanimity if $f(x, x, \ldots, x) = x \forall x$.

[5] Generally there exist no social choice rules satisfying Chichilnisky's three axioms, cf. Chichilnisky [6, 7]. In our case they exist because we restrict the domain of preferences, see also Chichilnisky and Heal [9]. These rules include various "generalized median" rules, such as those of Moulin [18], which are extensions of the median rule by the inclusion of non-existent voters, and those of Barberà, Gul and Stacchetti [2], elegantly defined by left- and right-coalition systems.

The attractive properties of these LCD rules are bought at a high price: there are very few such rules. Formally, LCD rules are nowhere dense in the space of continuous functions. Straightforward games are therefore not robust.[6]

In addition to simplicity, our characterization has clear advantages over alternative descriptions of straightforward rules in terms of medians and phantom voters[7]. LCD rules can be extended naturally to infinite populations, for which medians are not well-defined (see Heal [15]). Another advantage is that it provides a basis for analyzing the incentive-compatibility properties of Rawlsian rules. These have been widely studied and have the property that (locally) one individual[8] is dictatorial, the person who is in the worst position, and the rule is constant with respect to the preferences of all others. Therefore Rawlsian rules are straightforward.

Our results extend also to Nash equilibrium strategies. We show that being LCD is necessary and sufficient for truthful revelation to be a Nash equilibrium. So the apparently less demanding concept of Nash implementation in fact brings little in the way of greater generality.

We work with generalizations of single-peaked preferences,[9] in our case the indifference curves are families of ellipsoids. Choice spaces are linear subspaces of a Euclidean space.[10] The messages or strategies of the players are statements of their characteristics: these are either vectors in R^{n+} (bliss points of the single peaked preferences), or alternatively, preferences over R^n. Outcomes, or payoffs, are vectors in R^n. Each player seeks through strategic behavior to attain an outcome as close as possible to his or her optimal outcome or bliss point, according to some distance on R^n.

The paper is organized as follows: the following section introduces the results and provides geometric examples. Section three proves rigorously the results on straightforwardness, and section four does likewise for Nash implementation with separable regular games.[11] The main part of the paper uses only geometric arguments; longer proofs are in the Appendix.

2. The geometry of implementation

This section gives an introduction to the subject by providing examples and simple geometric interpretations of the results.

---

[6] A related fact was noted by Guesnerie and Laffont [13] in a different framework.

[7] Moulin [18] studies straightforwardness in terms of generalized median rules. His results apply only to one-dimensional choice spaces. Border and Jordan [4] work with so-called "phantom voters". They study voting rules where the population of voters is enlarged by imaginary or phantom voters.

[8] Not always the same individual, but always the individual occupying the position of being worst off.

[9] The same framework has been used by Moulin, Barberà Gul and Stacchetti, Barberà et al. [3], Border and Jordan [4] and van der Stel [23]. For an excellent recent review of this literature see van der Stel [23].

[10] Unlike Barberà et al., who work with discrete sets of choices.

[11] Separable regular games are defined fully below: separability means that $g: \mathfrak{R}^{mk} \to \mathfrak{R}^m$, $g = (g_1, g_2, \ldots, g_m)$. Regularity is a rank condition on the derivative of the game form.

We start with games where the players' characteristics are real numbers; later we consider more general cases. There are $k \geq 2$ players. Each player wishes to achieve an outcome in the real line which is as close as possible (in $\mathfrak{R}$) to this or her true "bliss point". Preferences are therefore represented by utility functions that are symmetric around a maximum value in $\mathfrak{R}$, the "bliss point". S is the space of strategies and A the space of outcomes. A game form $g: S^k \to A$ (also called a "rule") is a function which associates with each k-tuple of agents' strategies an outcome in A. A game g respects unanimity if $g(p_1, \ldots, p_k) = \bar{y} \in A$ when for all $i = 1, \ldots, k$ the preferences p_i, have the same bliss point $\bar{y}$. The game g just defined is called *straightforward* if the announcement of one's true characteristic is always a dominant strategy for each player.[12]

There is an equivalent expression for straightforward games, which we present here for clarity but which is unnecessary otherwise; one says that a game "implements" a social choice rule if the equilibria of the game are the outcomes of the social choice function applied to "true" individual preferences. Thus a straightforward game implements its game form g as a social choice function. The notion of equilibrium can be based on dominant strategies, or be a Nash equilibrium: both are considered in this paper.

A game which is not straightforward is called manipulable: in such games players have incentives to lie.

2.1. Manipulable rules

Standard games, such as average rules, are manipulable. It will help the intuition to see why. Consider the game as defined above, where

$$g: [0, 1]^2 \to [0, 1], \quad g(r_1, r_2) = \lambda r_1 + (1 - \lambda) r_2, \ \lambda \in [0, 1].$$

Figure 1 represents this game form: the slanted lines represent the hypersurfaces of the game form function g, $g^{-1}(r) = \{(r_1, r_2): g(r_1, r_2) = r\}$. The horizontal axis of the square are the strategies of player one; the vertical of player 2.

This game has an interesting characteristic: for any strategy s_2 of player two within the segment S, there exists a strategy for player one denoted $r(s_2)$, which can attain his/her preferred outcome or "bliss point" r_1, i.e., $g(r(s_2), s_2) = r_1$. It suffices to choose $r(s_2)$ so that $(r(s_2), s_2 \in g^{-1}(r)$. Furthermore, this optimal strategy for player one, $r(s_2)$, varies with s_2. Therefore, stating the true characteristic r_1 is generally *not* the best strategy for player one. In fact, it is easy to check that in general *this game has no dominant strategies.* This game is manipulable.

It is clear from the above discussion that, *to avoid manipulability, one must require that the optimal response* $r(s_2)$ *does not vary locally with* s_2. This implies, in the diagram of the game g, that the hypersurfaces $g^{-1}(r)$ are either (i) vertical, in which case $r(s)$ is always the same as s varies within a neighborhood, or (ii) horizontal with $r(s) \equiv s$, so that r cannot affect the outcome and s has no incentive to lie, or else (iii) that the game g has large indifference

---

[12] A strategy r is *dominant* for player one if for all s in S $g(r, s) = \max_{t \in S}(g(t, s))$, according to player one's preference.

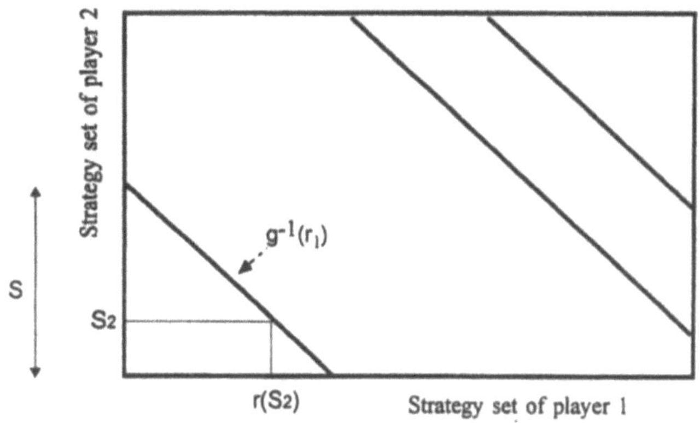

Fig. 1. The game $g(r_1, r_2) = \lambda r_1 + (1 - \lambda) r_2$, $\lambda \in [0, 1]$

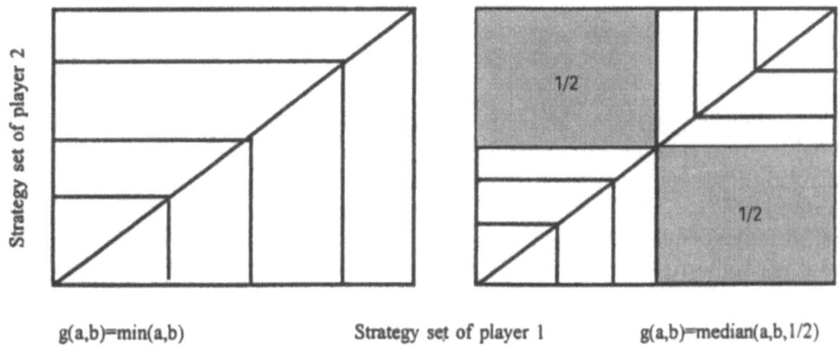

Fig. 2. Games whose level sets are horizontal or vertical are straightforward

surfaces so that both r and s remain constant locally with changes in the strategies s. Examples of continuous games of this sort are in Fig. 2. It can be verified that both of these games are indeed straightforward.

The next section proves rigorously that games such as those represented in Fig. 2, are always straightforward. Furthermore, the results of next section establish that all straightforward games are of this form. Why?

2.2. Illustrating the results

Why should straightforward rules be LCD? An intuitive argument is as follows.

Consider a game as above, $g : \mathfrak{R}^2 \to \mathfrak{R}$. Let g be onto and straightforward. Somewhat surprisingly, in this case, g must be continuous (see Theorem 1 below). Define now the manipulation set M_{r_2}: it is the set of outcomes which the first player can achieve when player two plays r_2. Then if the "true" bliss point of player 1, r_1, is in the interior of M_{r_2}, denoted $\mathring{M}_{r_2}$, r_1 is by definition achievable by player one with true characteristic r_1, by straightforwardness.

Fig. 3. 2 players with strategies & outcomes in $\Re$. For $r_1 \in$ interior M_{r_2}, g is a projection: otherwise it is constant

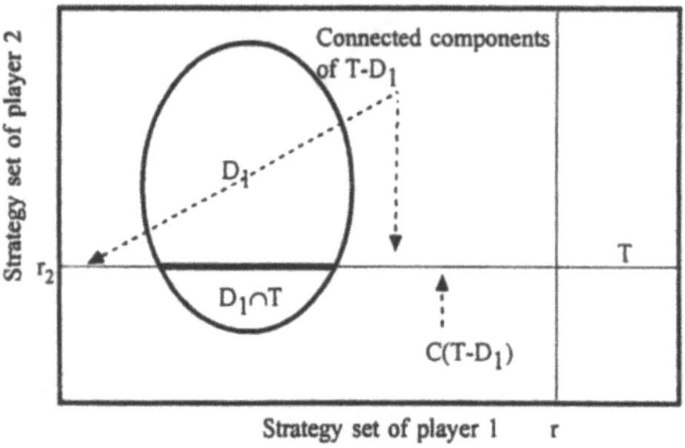

Fig. 4. The game is LCD, and D_1 is the region in which 1 is dictatorial

This implies $g(r_1, r_2) = r_1$. Since the condition $r_1 \in M\mathring{M}_{r_2}$ is satisfied in an open neighborhood of (r_1, r_2) by continuity, it follows that $g(r_1, r_2) \equiv r_1$ in a neighborhood of (r_1, r_2), when $r_1 \in M\mathring{M}_{r_2}$. Thus, g is here a projection locally. Now assume $r_1 \notin M_{r_2}$. Then $g(r_1, r_2)$ must be the closest point to r_1 within M_{r_2} by straightforwardness. Clearly, as r_1 varies within a neighborhood, this outcome remains constant, see Fig. 3. If $r_2 \in M_{r_2}$ the same argument applies, so that the outcome remains locally constant as r_2 varies locally as well. Therefore in this second case $g(r_1, r_2)$ is locally a constant map.

Th e remaining case is when either r_1 or r_2 is in the boundary of M_{r_2} and this occurs on a set of points $(r_1, r_2) \in \Re^2$ of measure zero. Therefore, a.e. a straightforward onto game is locally constant or dictatorial. We have therefore shown that a straightforward game must be LCD.

The converse is also easy to visualize. Assume g is LCD. Let D_1 be the subset of $\Re^2$ where player 1 is dictatorial, i.e., $g(r_1, r_2) \equiv r_1$. D_1 can be shown to be a connected set.

If $(r_1, r_2) \in D_1$, then r_1 is clearly the best strategy for player one with true characteristic r_1. Otherwise, if $r_1 \notin D_1$, let $T = \{(r, s) \in R^2 \colon s = r_2\}$ and $T - D_1$ be the part of T not in D_1. By assumption, g is locally constant on $T - D_1$ with respect to its first coordinate; since g is continuous, g must be constant on any connected component of $T - D_1$, $C(T - D_1)$. Assume that player one's true preference is r_1 and $(r_1, r_2) \in T - D_1$, see figure 4. Any point in this component of $T - D_1$ therefore gives the same outcome as (r_1, r_2) so that there are no incentives to lie within this component of $T - D_1$. Furthermore, by continuity, $g(r_1, r_2) = r_1$ if[13] $(r_1, r_2) \in \partial \mathring{C}(T - D_1)$. In addition, the strategy $r' \neq r_1$ is also less preferable to r_1 if $r' \in \mathring{D}_1$ because $g(r, r_2) = r$ and is therefore further away from r_1 than is $g(r_1, r_2)$. Finally, if (r', r_2) is in another connected component of $T - D_1$ where g is locally constant, see figure 4,

[13] The boundary of a set X is denoted ∂X.

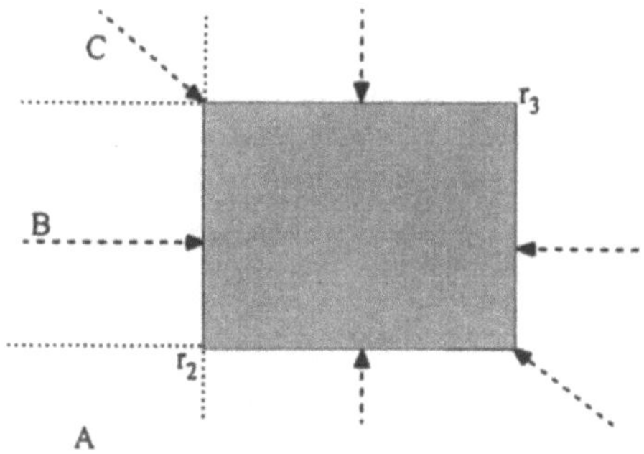

Fig. 5. Outcomes are 2-dimensional and there are 3 agents. Agents 2 and 3 announce r_2 and r_3: if agent 1 announces in the shaded area, 1 determines the median in each coordinate and is a dictator. For announcements by 1 outside the shaded area, the rule acts as a projection onto this area

$g(r', r_2)$ is still further away from r_1 than is $g(r_1, r_2)$ because it is at least as far as $g(r', r_2)$ where $r' \in \partial C(T - D_1)$. Therefore a rule which is LCD and onto is straightforward, as we wished to show.

Up to now the player's characteristics are real numbers. Now we consider two higher dimensional examples.

Example 1. Let $n = 2$ so that choices and bliss points are in $\Re^2$ and let the number of players $k = 2$. Define $g(r_1, r_2) = (r_{11}, r_{22})$, where $r_1 = (r_{11}, r_{12})$ and $r_2 = (r_{21}, r_{22})$.

Thus agent 1 is dictatorial in the first component, and agent 2 in the second. Clearly the rule g is locally constant or dictatorial, and is straightforward. Agent 2's manipulation set is a vertical straight line through 1's announcement, and 1's is a horizontal straight line through 2's, and any announcement by 2 (or 1) leads to an outcome which is the horizontal (or vertical) projection of this into the vertical (or horizontal) line through 1's (or 2's) announcement.

Example 2. Now let $n = 2$ and $k = 3$ and $g(r_1, r_2, r_3) = (x_1, x_2, x_3)$ where $x_j = \text{median } (r_{1j}, r_{2j}, r_{3j})$.

This is a coordinate-wise median rule. Fig. 5 shows the manipulation set of agent 1, when 2 and 3 have announced r_2 and r_3 respectively. The manipulation set is shaded. If 1's announcement is contained in this, it is the median in both components and 1 is a dictator.

Consider regions A and B as indicated in Fig. 5. If r_1 is in region B, then r_1 has the median vertical component and r_2 the median horizontal component and the outcome is (r_{21}, r_{12}). Hence in region B, $g(., r_{-i})$ acts a horizontal projection onto the manipulation set, where r_i is the vector r with the i-th component deleted. In region A, r_2 has the median in both components and the outcome is r_2. Hence in A g acts to project to the nearest point of manipulation set, r_2. In region C, r_2 has the horizontal and r_3 the vertical median, so the outcome is (r_{21}, r_{32}) and all points in C are mapped to the nearest corner of the shaded set. It is now routine to verify that $g(., r_{-i})$ acts elsewhere as shown in Fig. 5, which illustrates its action as a projection onto a convex set bounded by coordinate hyperplanes.

3. Main results

In the previous section we gave intuitive arguments about the equivalence of straightforward rules and rules which are LCD. We now give a formal and general statement of that result and of other related results.

3.1. Notation and definitions

Let X be the choice space, $X = \mathfrak{R}^{n+}$. A *preference* p_i over X is given by two objects: a "bliss" point y^i in X, and a distance function d_i: a choice x is preferred to another z if x is closer than z to the bliss point y^i, i.e., $d_i(x, y^i) < d_i(z, y^i)$. The distance $d_i(x, y)$ is given by $\sum_{j=1}^{n} m_j(x_j - y_j)$, where (m_j) is a strictly positive vector in $\mathfrak{R}^n$ (i.e., d is not degenerate). The indifference surfaces of p_i are then convex ellipsoids with center at the bliss point y^i and axes parallel to the coordinate axes.

The *space of strategies or messages* S is either (i) $S = \mathfrak{R}^{n+}$, in which case each message in $\mathfrak{R}^{n+}$ is interpreted as a statement of an agent's preferred outcome, or (ii) $S = P$, where P is the space of all preferences (distances and bliss points) defined above. Thus either $(\mathfrak{R}^n)^k$ or P^k is the space S^k of strategy profiles for k players. Since a preference in P is uniquely identified by its bliss point and its metric[14], $P \approx \mathfrak{R}^{2n+}$. The space of outcomes A is $\mathfrak{R}^{n+}$ in either case.

A *game form* is now a map $g: S^k \to A$. Continuity of g is defined with respect to the usual topology of Euclidean spaces. When the game form g can in principle take any value in $\mathfrak{R}^n$, g is called onto.

A *game* is given by a game form as above, and a family $\{p_i\}$ of preferences over outcomes, designated by matrices $M_i \in P$, $i = 1, \ldots, k$.

The symbol (m_i, m_{-i}) denotes a message or strategy profile in S^k, with its i-th. component equal to m_i in S and where m_{-i} is a $k-1$ vector of strategies for players other than k.

A strategy profile $(m_1^*, \ldots, m_k^*)$ is a *dominant strategy equilibrium* if for all $i = 1, \ldots, k$ and m_{-i}, the outcome $g(m_1^*, m_{-i})$ is preferred to the outcome $g(m_i, m_{-i})$, for all $m_i \in P$, according to player i's preference p_i.

---

[14] The inner product $\langle , \rangle_i$ defining the metric d_i can be identified by a matrix M, such that $\langle x, y \rangle_i = \langle Mx, y \rangle$, where $\langle , \rangle$ denotes the standard inner product in $\mathfrak{R}^n$. The matrix M is symmetric and also positive definite (in order to be non-degenerate). We shall assume that preferences are separable, in the sense that if any two choices x and z differ only on their k-th coordinates, x is preferred to z when the coordinate x_k is preferred to the coordinate z_k (i.e., x_k is closer to the k-th coordinate of the bliss point y than z_k). In this case the matrix M is diagonal, and we can write

$$d_i(x, y) = \|x - y\|_i = \langle m, (x - y) \rangle = \sum_{j=1}^{n} m_j(x_j - y_j),$$

where m is a positive vector in $\mathfrak{R}^n$. We assume m is strictly positive, since otherwise the metric would be degenerate. A preference is thus completely identified by a bliss point y_i in $\mathfrak{R}^{n+}$, and by a strictly positive vector $m \in \mathfrak{R}^n$.

A game g is *straightforward* if $(r_1, \ldots, r_k)$ is a dominant strategy equilibrium for players with characteristics $(r_1, \ldots, r_k)$ in S^k, i.e., truthful messages about characterstics are dominant strategies for each player.

The *pre-manipulation set* of m_{-i} is the set

$$N_{m_{-i}} = \{(m_i, m_{-i}) \in S^k \colon m_i \in S\}.$$

A function $f: \mathfrak{R}^s \to \mathfrak{R}$ is *locally constant or dictatorial* (*LCD*) if it is continuous, and for almost all[15] x in $\mathfrak{R}^s$, there exists a neighborhood $N_x \subset \mathfrak{R}^s$ with

$$f/N_x = \text{constant}$$

or

$$f/N_x(y) \equiv y_d, \quad \text{for some } d \in \{1, \ldots, k\}, \text{ for all } y \text{ in } N_x.$$

For higher dimensional domains and ranges, a function $f: (\mathfrak{R}^m)^k \to \mathfrak{R}^m$ is called *separable* if the j-th. coordinate of the image depends only on the j-th. coordinates of the arguments, i.e.,

$$f(x_1^1, \ldots, x_m^1, \ldots, x_1^k, \ldots, x_m^k) = f_1(x_1^1, \ldots, x_1^k), \ldots, f_m(x_m^1, \ldots, x_m^k).$$

A function $f: (\mathfrak{R}^m)^k \to \mathfrak{R}^m$ is called *LCD coordinate-wise or LCD*, if it is separable, $f = (f_1, \ldots, f_m)$, and if each f_i is LCD for $i = 1, \ldots, m$.

We consider here game forms which are not necessarily onto: their images are *linear subsets* of $A = \mathfrak{R}^n$, i.e., $g(S^k) = \mathfrak{R}^s \subset \mathfrak{R}^n$ with $s \leq n$, or $g(S^k) = \{(x_1, \ldots, x_s) \in \mathfrak{R}^s \colon a_i \leq x_i \leq b_i, \text{ each } i\}$. Such games do not necessarily respect unanimity even if they are straightforward. The next result will show that any such straightforward game $g: S^k \to A$ is LCD, where the strategy space S is either $\mathfrak{R}^n$ or the space of preferences P defined above. Furthermore, being LCD is also sufficient for straightforwardness.

Let X be a subset of a topological space Y. Then X is *residual* if it is a countable intersection of open and dense subsets of Y. A residual set in a complete normed space is always dense.

3.2. Lemmas

The following are simple but useful properties of straightforward games.

Lemma 1. *A straightforward game $g: (\mathfrak{R}^n)^k \to \mathfrak{R}^n$ respects unanimity if and only if it is onto.*

Proof. In the Appendix. ■

Lemma 2. *If a game $g: \mathfrak{R}^2 \to \mathfrak{R}$ is straightforward and onto, then the outcome $g(r_1, r_2)$ is contained in the segment $[r_1, r_2]$.*

Proof. In the Appendix. ■

Lemma 3. *If $k = 2$, g is straightforward and its image is a segment $[a, b] \subset \mathfrak{R}$, then either the outcome $g(r_1, r_2)$ is in the segment $[r_1, r_2]$, or else $g(r_1, r_2)$ is in the*

[15] "*Almost all*" indicates for all points in $\mathfrak{R}^s$, except possibly on a subset of Lebesgue measure zero in $\mathfrak{R}^s$.

boundary of the segment $[a, b]$ *denoted* $\partial[a, b]$. *Furthermore, if the strategy* r_1 *is not in* $[a, b]$, *then the outcome* $g(r_1, r_2)$ *is the same as the outcome* $g(x, r_2)$, *where* x *is the closest point to* r_1 *in* $[a, b]$.

Proof. In the Appendix. ■

Now define the *manipulation set* M_{r-i}, i.e., the set of outcomes that can be obtained by player i when all other players have announced a vector of messages $(r_1, \dots, r_{i-1}, r_{i+1}, \dots, r_k) = r_{-i}$ in $\Re^{k-1}$. $M_{r-i} = \{y\colon y = \phi(r, r_{-i}), r \in \Re\}$

Lemma 4. *If* $g: \Re^k \to \Re$ *is straightforward and the strategy* r_i *is in* $M_{r_{-i}}$ *then the outcome* $g(r_1, \dots, r_k) = r_i$. *If* $r_i \notin M_{r_{-i}}$, *then the outcome* $g(r_1, r_k)$ *is the closest point to* r_i *in the boundary of* $M_{r_{-i}}$, *denoted* $\partial M_{r_{-i}}$. In particular, $g(\Re^n)^k$ is closed.

Proof. In the Appendix. ■

3.3. Straightforward games with a convex image are continuous

Theorem 1. *If the choice space is one-dimensional,* $g: \Re^k \to \Re$ *is straightforward and its image* $g(\Re^k)$ *is convex, then* g *is continuous. In particular, if* g *is straightforward and respects unanimity, then is it continuous.*

Proof. In the Appendix. ■

Remark 1. Not all straightforward games are continuous. Figure 6 gives an example of a discontinuous straightforward game with a non-convex range.

It is clear that the game in Fig. 6a is straightforward for any pair (r_1, r_2) in the interior of one of the shaded areas. When the characteristics of the players (r_1, r_2) are either $r_1 = \frac{1}{2}$ or $r_2 = \frac{1}{2}$, it is easy to see that g is straightforward also, since the outcomes $\frac{1}{4}$ and $\frac{3}{4}$ are equi-distant from $\frac{1}{2}$. Saying the truth is thus a dominant strategy for both players. The example in Fig. 6a can be generalized to produce straightforward games with a large, even countably infinite, number of discontinuities. Figure 6b shows a rule which is LCD, discontinuous and *not* straightforward. So without continuity the equivalence of straightforwardness and being LCD does not hold, although straightforwardness and a convex range together imply continuity (Theorem 1).

3.4. Straightforward rules and LCD rules

Lemma 5. *Let* $\phi: P^k \to A$ *be a locally constant or dictatorial (LCD) rule. Then* ϕ *is straightforward.*

Proof. The proof is in the Appendix. ■

We can now state formally the main result of this paper:

Theorem 2. *Let* $g: S^k \to A$ *be a game form. Then* g *is straightforward if and only if* g *is locally constant or dictatorial.*

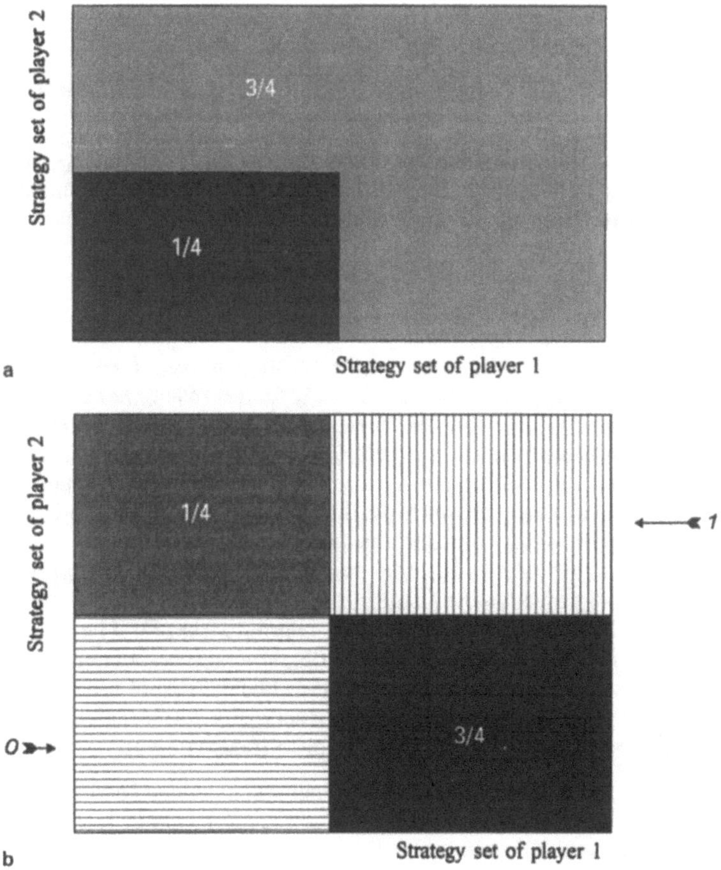

Fig. 6. a The game $g:[0,1]^2 \to [0,1]$ is defined by $g(r_1, r_2) = 1/4$ if $r_1 \leq 1/2$ and $r_2 \leq 1/2$; $g(r_1, r_2) = \frac{3}{4}$ otherwise. **b** In this case, $g(x_1, x_2) = 0$ for $0 \leq x_1, x_2 \leq \frac{1}{2}$, $=1$ for $\frac{1}{2} \leq x_1, x_2 \leq 1$, $=\frac{1}{4}$ for $0 < x_1 \leq \frac{1}{2}$ and $\frac{1}{2} \leq x_2 \leq 1$, and $=\frac{3}{4}$ for $\frac{1}{2} < x_1 \leq 1$ and $0 \leq x_2 < \frac{1}{2}$

Proof. The proof is in the Appendix. Sufficiency is clearly established by Lemma 5 above. The formal proof builds on the intuitive arguments of the previous section. ■

An immediate implication of our main result is that smooth straightforward rules onto $\mathfrak{R}^n$ are coordinate-wise dictatorial:

Corollary 1. *If $g:(\mathfrak{R}^n)^k \to \mathfrak{R}^n$ is straightforward and differentiable, and its image is $\mathfrak{R}^n$, then g is coordinate-wise dictatorial. In particular, g is not anonymous.*

Proof. Since by Theorem 1, g is LCD, by differentiability each coordinate must be either constant everywhere, or a projection. If the map g was constant in one coordinate, g could not be onto $\mathfrak{R}^n$. Therefore g must be dictatorial coordinate-wise, implying that each coordinate j is identically equal to the

j-th. coordinate of some (fixed) player, d_j. In this case a permutation of the agents alters the outcome. This proves the corollary. ■

Corollary 2. *Let $\phi: P^k \to A$ be a Rawlsian social choice rule. Then ϕ is straightforward.*

Proof. Rawlsian rules are those which maximize the utility of one agent – the agent who is worst off. This agent is therefore dictatorial. The identity of this agent may vary, so that the rule s locally dictatorial. ■

3.5. *Robustness of straightforward games*

We shall now study the robustness of straightforward games. Consider the family of all straightforward game forms $g: P^k \to A$, where A is linear subset of $\mathfrak{R}^n$. As we have shown in the proof of Theorem 2, each such g is always representable as a continuous function $g: (\mathfrak{R}^n)^k \to A$. This is because straightforwardness implies that the outcome that g assigns to messages in P^k depends only on the bliss points of their messages so that Theorem 1 applies and g is continuous. We shall therefore consider the family of all continuous maps $f: (\mathfrak{R}^n)^k \to A$, denoted $C^0((\mathfrak{R}^n)^k, A)$. In order to give this space an appropriate norm topology, we shall consider the space A to be bounded, i.e., A is a cube in $\mathfrak{R}^n$, denoted I^n. The message space is also taken to the I^n. The space $C^0((\mathfrak{R}^n)^k, A)$ is given the sup norm topology, with the distance between two functions f and g given by $\sup_{x \in \mathfrak{R}^{nk}} \|f(x) - g(x)\|$. The next result establishes that the property of straightforwardness is not structurally stable, since small deformations of a straightforward function are not straightforward.

Theorem 3. *The family of non-straightforward games on a bounded choice space is a residual set of the space of continuous maps ($C^0(I^{nk}, I^n)$) from $I^{n,k}$ to I^k, and, in particular, is a dense set.*

Proof. The proof is in the Appendix.

4. Nash implementation

The following result analyzes the problem of Nash implementation in cases where preferences P, messages M and outcomes A are one dimensional. Consider a game form g: $M^k \to A$, where both the message space M and the outcome space A are one dimensional, $M = A = I$, I the unit interval in $\mathfrak{R}$. We shall require g to be a C^{k+1} map,[16] and to satisfy a *regularity condition* (1) defined as follows.

Regularity condition: Let $\eta = (\eta_1, \ldots, \eta_{k-j})$ denote a non-empty subset of $k - j$ integers in $\{1, \ldots, k\}$. Define the map $g_\eta: M^k \times I \to \mathfrak{R}^{k-j+1}$ by

$$g_\eta(m, b) = (Dg_{\eta 1}(m) + g(m), \ldots, Dg_{\eta k-j}(m) + g(m), g(m))$$

[16] For related results on implementation of smooth maps, see Laffont and Maskin [17].

for all $(m, b) \in M^k \times I$ where $Dg_{\eta 1}$ is the matrix of first partial derivatives of g with respect to η_1. Let Δ_η be the "diagonal" subset

$$\Delta_\eta = \{(x_1, \dots, x_{k-j+1}) \in \Re^{k-j+1}: x_i = x_j \forall i = 1, \dots, k-j+1\}.$$

Then g must be transversal to $\Delta\eta$, i.e.

$$g_\eta \pitchfork \Delta_\eta, \ \forall \eta \subset \{1, \dots, k\}. \tag{1}$$

Condition (1) applies to any $(m, b) \in M^k \times I$ such that all $k-j+1$ coordinates of its image in R^{k-j+1} under $g_\eta(m, b)$ are equal. This is equivalent to $g(m) = b$ and $Dg_{\eta_1}(m) + b = b, \dots, Dg_{\eta_{k-j}}(m) + b = b$, or equivalently $g(m) = b$ and $Dg_{\eta_1}(m) = \dots = Dg_{\eta_{k-j}}(m) = 0$. In this case condition (1) implies that the gradient $Dg_\eta(m, b)$ has rank $k-j$ at such points, i.e., at points $(m, b) \in M^k \times I$ mapping into the diagonal of R^{k-j+1}.

A game form g satisfying (1) is called *regular*. Note that condition (1) does not imply the gradient $Dg(x) \neq 0$ for all $x \in I^k$. The following lemma establishes that (1) is satisfied for a generic set of C^{k+1} games:

Lemma 6. *Consider the family G of all C^{k+1} maps $\{g: I^k \to I\}$, endowed with the C^{k+1} sup topology. Condition (1) is satisfied on a residual (and in particular dense) set of G.*

Proof. The proof is omitted due to space constraints, but can be obtained from the authors. It is in a Columbia Business School Working Paper with the same name as this article.

Note that the genericity of the condition (1) on C^{k+1} games, does not imply that the set of rules to be Nash implemented is a generic set. The following theorem proves that only very special rules will be Nash-implementable. The reason for this is that a large class of regular games will implement the same rule.

Theorem 4. *Let ϕ be a continuous social choice rule, $\phi: P^k \to A$ where $P = A = I$, the unit interval in $\Re$, and $k \geq 2$[17]. Let $M = I$ be the message space consisting of statements on bliss points of individual preferences. If the rule ϕ is Nash implementable by a regular game $g: M^k \to A$, then ϕ is locally constant or dictatorial (LCD).*

Proof. This is proved in the Appendix. ■

From Theorem 4 we obtain the following result, which is valid for any euclidean space of outcome or message spaces:

Theorem 5. (*1*) *Let $\phi: P^k \to A$ be a rule which is Nash implementable by a separable regular game*[18] *g. Then ϕ is straightforward and LCD.* (*2*) *If*

---

[17] The space of preferences P is here identified with the space of outcomes $A = I$ by assigning to each preference $p \in P$ its (unique) bliss point b in I. Therefore the space P is given the standard topology of I in which two preferences p_1, p_2 close it their bliss points b_1, b_2 are close in I. Continuity of ϕ refers here to this topology on P.

[18] A separable game $g: I^{km} \to I^m$, $g = (g_1, \dots, g_m)$ is said to be regular if each component function g_i, $i = 1, \dots, m$, is regular.

$\phi: P^k \to A$ is straightforward and regular, then ϕ is LCD and (of course) Nash implementable by a separable regular game.

Proof. First note that if a rule ϕ is Nash implementable by a separable game g, then ϕ is separable. Therefore by Theorem 4 if ϕ is Nash implementable by a separable regular game $g = (g_1, \ldots, g_m)$, then ϕ is separable, i.e., $\phi = (\phi_1, \ldots, \phi_m)$, and each ϕ_i is implementable by a regular game. Therefore, each ϕ_i is LCD, which implies, by definition, that ϕ is LCD.

The converse is immediate: any straightforward rule ϕ is obviously Nash implementable by the game form it defines. Therefore we can apply Theorem 2 and this proves that ϕ must be locally constant or dictatorial. This completes the proof. ■

5. Conclusions

Being locally constant or dictatorial has been shown to be a necessary and sufficient condition for straightforwardness. This is an intuitively appealing result: it is clear that constant rules or dictatorial rules are straightforward. Our result says that the only straightforward rules are those obtained by "patching together" in a continuous fashion rules where are locally constant or locally dictatorial in each coordinate. However, such rules may be quite complex: many LCD games are simultaneously continuous and anonymous, and also respect unanimity. Hence they satisfy the axioms introduced by Chichilnisky [6], and subsequently used by others, for characterizing ethically acceptable social choice rules. Such axioms are not satisfied in general without restrictions on preferences, see, e.g., Chichilnisky [6] and Chichilnisky and Heal [9].

Moreover, as the "patching together" of constant and dictatorial rules can be done continuously but not smoothly, it immediately follows that if one requires smoothness, the rules must be either constant or dictatorial on each component. If onto, the rule must be dictatorial coordinate-wise, and thus cannot be anonymous.

We therefore have a rather simple and intuitive characterization of the possibilities for straightforward implementation, which shows that only a rather special type of rule is straightforward, and that the class of such rules is not robust. We have also shown that being locally constant or dictatorial a necessary and sufficient condition for implementability via the Nash equilibria of a separable regular game. So with separable rules, even though Nash implementable rules is much less demanding, the relaxation of the implementation concept (from straightforwardness to Nash) does not change the results.

A. Appendix

A.1. Proofs of results on straightforwardness

A.1.1. Proofs of lemmas

Lemma 1. *A straightforward game $g: (\Re^n)^k \to \Re^n$ respects unanimity if and only if it is onto.*

Proof. If g respects unanimity then the image of g covers $\mathfrak{R}^n$. The converse is also immediate. Assume g is straightforward and onto; for any $r \in \mathfrak{R}^n$ let $r = g(r_1, \ldots, r_k)$, i.e., r is attainable by player 1 (by announcing r_1) when the other players announce a $k-1$ vector $(r_2, \ldots, r_k)$. Since r is the best outcome for player 1 with true characteristic r, it follows by straightforwardness that by stating the truth, player 1 must be able to attain r, i.e., $g(r_1, \ldots, r_k) = r \Rightarrow g(r, r_2, \ldots, r_k) = r$.

Iterating this procedure, one obtains $g(r, \ldots, r) = r$, i.e., g respects unanimity. ■

Lemma 2. *If a game $g: \mathfrak{R}^2 \to \mathfrak{R}$ is straightforward and onto, then the outcome $g(r_1, r_2)$ is contained in the segment $[r_1, r_2]$.*

Proof. Being onto and straightforward, g respects unanimity by lemma 1, so that $g(r_2, r_2) = r_2$. It follows that for any r_1 $g(r_1, r_2)$ must be at least as preferable to player one with true characteristic r_1 as is r_2, so $g(r_1, r_2)$ is closer to r_1 than is r_2. Since this is also true for player 2, then $g(r_1, r_2) \in [r_1, r_2]$. ■

Lemma 3. *If $k = 2$, and the game form g is defined on $\mathfrak{R}$, is straightforward and its image is a segment $[a, b] \subset \mathfrak{R}$, then either the outcome $g(r_1, r_2)$ is in the segment $[r_1, r_2]$, or else $g(r_1, r_2)$ is in the boundary of the segment $[a, b]$ denoted $\partial[a, b]$. Furthermore, if the strategy r_1 is not in $[a, b]$, then the outcome $g(r_1, r_2)$ is the same as the outcome $g(x, r_2)$, where x is the closest point to r_1 in $[a, b]$.*

Proof. Consider first the case when $[r_1, r_2] \cap [a, b] = \phi$. In this case by straightforwardness the outcome of (r_1, r_2) must be the closest point to r_1 and to r_2 in $[a, b]$, i.e., a point in $\partial[a, b]$. Secondly, consider the case of $[a, b] \subset [r_1, r_2]$. Then the lemma is obviously true, by the assumptions on g. Thirdly, note that if $r_1 \in [a, b]$ and $r_2 \in [a, b]$, then obviously $g(r_1, r_2) \in [r_1, r_2]$. Finally, suppose that both r_1 and r_2 are in $[a, b]$. Then assume $g(r_1, r_2) = r$. By straightforwardness, $g(r, r_2) = r$ and similarly $g(r, r) = r$. Hence the restriction of g on $[a, b]$, $g/[a, b]$, is onto and lemma 2 applies. ■

Now recall the definition of the *manipulation set* M_{r-i}, i.e., the set of outcomes that can be obtained by player i when all other players have announced a vector of messages $(r_1, \ldots, r_{i-1}, r_{i+1}, \ldots, r_k) = r_{-i}$ in $\mathfrak{R}^{k-1}$, $M_{r-i} = \{y: y = \phi(r, r_{-i}), r \in \mathfrak{R}\}$.

Lemma 4. *If the game form $g: (\mathfrak{R}^n)^k \to \mathfrak{R}^n$ is straightforward and the strategy r_i is in M_{r-i} then the outcome $g(r_1, \ldots, r_k) = r_i$. If $r_i \in M_{r_{-i}}$, then the outcome $g(r_1, \ldots, r_k)$ is the closest point to r_i in the boundary of $M_{r_{-i}}$, denoted $\partial M_{r_{-i}}$.* In particular, $g(\mathfrak{R}^n)^k$ is closed.

Proof. This is immediate. If $r_i \in M_{r-i}$, then r_i is attainable by the i-th. player, given the other messages r_{-i}. Then by straightforwardness $g(r_1, \ldots, r_i, \ldots, r_k) = r_i$. If $r_i \notin M_{r_{-i}}$, then by straightforwardness $g(r_1, \ldots, r_i, \ldots, r_k)$ must be the best attainable by player i with characteristics r_i. Thus $g(r_1, \ldots, r_i, \ldots, r_k)$ must be the closest point to r_i in $M_{r_{-i}}$, and thus in $\partial M_{r_{-i}}$. ■

Theorem 1. *If the choice space is one-dimensional, $g: \mathfrak{R}^k \to \mathfrak{R}$ is straightforward and its image $g(\mathfrak{R}^k)$ convex, then g is continuous. In particular, if g is straightforward and respects unanimity, then it is continuous.*

Proof. The strategy of the proof is as follows. We consider first the case of 2 players, and then extend the result to any number by induction. In the case of two players, we deal first with the case in which neither r_1 nor r_2 are in the image of $\mathfrak{R}^2$ under g, $g(\mathfrak{R}^2)$. In the remaining case we show that the graph of g is closed, and hence that it is continuous.

Consider first the case of two players, $k=2$. Assume first that neither r_1 nor r_2 are in $g(\mathfrak{R}^2)$ and let the sequence $(r_1^h, r_2^h) \to (r_1, r_2)$. If $[r_1, r_2]$ does not intersect the image of g, then there exists N such that $[r_1^h, r_2^h]$ will not intersect $g(\mathfrak{R}^2)$ either $\forall h > N$. Thus, for $h > N$, $g(r_1^h, r_2^h) \equiv x$, where x is a point in $\partial g(\mathfrak{R}^2)$, by Lemma 3. Since the outcome $g(r_1, r_2)$ is the same point x in this case, this proves continuity.

If r_1 is not in the image $g(\mathfrak{R}^2)$ but r_2 is, then $g(r_1, r_2) = g(x, r_2)$ by lemma 3, for some x in $g(\mathfrak{R}^2)$, so we consider next the case where *both* r_1 *and* r_2 *are in the image* of g. We show that in this case the graph of g is closed.

Let r_1 and r_2 be in $g(\mathfrak{R}^2)$ and $r_1 \neq r_2$. By Lemma 4 if $r_1 \in M_{r_{-1}}$, the manipulation set of agent 1, then $g(r_1, r_2) = r_1$ and g is continuous. Now assume that $r_1 \notin M_{r_{-1}}$. Since $\lim_h r_i^h = r_i$, $i = 1, 2$, $\exists N$ such that for $h > N$, $r_1^h \neq r_2^h$, and

$$g(r_1^h, r_2^h) = a \in \partial M_{r_{-1}}.$$

Since $r_2^h \to r_2$, by lemma 4, $g(r_1^h, r_2) = a$ for $h > N$, as a will also be the nearest point in $M_{r_{-1}}$ to r_1^h. This implies that

$$\lim_h g(r_1^h, r_2) = a.$$

We now claim that both points a and $g(r_1, r_2)$ must be at the same distance from r_1. To see this note that if a is nearer then player one could obtain a better outcome by stating r_1^h, whereas if $g(r_1, r_2)$ is nearer then player 1 with true preference r_1^h has an incentive to misrepresent and state r_1 as her true preference. So either $g(r_1, r_2) = a$ or the two points $g(r_1, r_2)$ and a are equidistant from r_1 on opposite sides. However, $g(r_1, r_2) \in [r_1, r_2]$ and $g(r_1^h, r_2) \in [r_1^h, r_2]$ by Lemma 3, implying that $g(r_1, r_2)$ and a cannot be on opposite sides of r_1 so that

$$g(r_1, r_2) = \lim_h g(r_1^h, r_2) = a.$$

Since $\lim_h (r_1^h, r_2) = \lim_h g(r_1^h, r_2^h) = g(r_1, r_2)$, we have shown that the limits of points in the graph of g, i.e., points of the form $\lim_h (r_1^h, r_2^h, g(r_1^h, r_2^h))$, are always in the graph of g, since they are equal to $(r_1, r_2, g(r_1, r_2))$. Thus, the map g has a *closed graph*, which is equivalent to being continuous. This completes the case in which both r_1 and r_2 are in the image of g and $r_1 \neq r_2$.

Now we consider the case where $r_1 = r_2$. First assume that $r_1 = r_2$, and they are both in the *interior* of the image $g(\mathfrak{R}^2)$. Then since $g(r_1^h, r_2^h) \in [r_1^h, r_2^h]$ for $h > N$, by lemma 3, we can pick sequences r_1^h and r_2^h such that without loss of generality $r_1^h < r_1 = r_2 < r_2^h$ so that $\lim_h g(r_1^h, r_2^h) = r_1 = g(r_1, r_2)$ and continuity is again ensured.

If $r_1 = r_2$ and they are both in the *boundary* of $g(\mathfrak{R}^2)$ then for $h > N$ we can assume without loss of generality one of three cases:

either $r_1^h \in \partial g(\mathfrak{R}^2)$ and $r_2^h \in g(\mathfrak{R}^2)$ in which case $g(r_1^h, r_2^h) = r_1^h$ by lemma 3; or both r_1^h and r_2^h are not in $g(\mathfrak{R}^2)$, in which case by lemma 3

$$g(r_1^h, r_2^h) = r_1$$

or, finally, both r_1^h and r_2^h are in $g(\mathfrak{R}^2)$, in which case

$$g(r_1^h, r_2^h) \in [r_1^h, r_2^h].$$

In any of these three cases, $\lim_h g(r_1^h, r_2^h)$ is $g(r_1, r_2)$ so that as before, the graph of g is closed, and thus g is continuous. This completes the proof of continuity for two players.

The argument for two players is clearly valid when there are $k > 2$ players, provided one is restricted to sequences in $\mathfrak{R}^k$ in which only two (the i-th. and j-th.) players vary their messages, i.e., sequences of the form

$$\lim(r_1, \ldots, r_j^h, r_{j+1}, \ldots, r_i^h, r_{i+1}, \ldots, r_k) = (r_1, \ldots, r_k) \in \mathfrak{R}^k.$$

We shall now prove continuity of g by induction, assuming continuity when up to $k - 1$ players are allowed to vary their messages.

Inductive hypothesis: g is continuous in any $k - 1$ of its arguments.

Let $(r_1^h, \ldots, r_k^h) \to (r_1, \ldots, r_k)$, and denote by r^h the value $g(r_1^h, \ldots, r_k^h) = r^h$. If for all i, r_i^h is in player i's manipulation set, i.e., $r_i^h \in M_{r_{-i}}^h$, then all r_i^h's must be identical, since in this case

$$g(r_1^h, \ldots, r_k^h) = r_i^h \quad \text{for all } i,$$

by Lemma 4. Continuity is assured in this case, since $r_i^h \to r_i$ for all i, and $g(r_1, \ldots, r_k) = r_i \ \forall i$. Otherwise, if some message r_i^h is *not* in $M_{r_{-i}}^h$, then by Lemma 4, for $h > N$

$$r^h = g(r_1^h, \ldots, r_{i-1}^h, r_i, r_{i+1}^h, \ldots, r_k^h) = g(r_1^h, \ldots, r_k^h),$$

because by choosing N sufficiently large we can ensure that r_i and r_i^h will be as close as desired. The problem is therefore reduced to one in which only $k - 1$ messages are allowed to vary, and by the induction hypothesis this completes the proof. ∎

Lemma 5. *Let $\phi: P^k \to A$ be a locally constant or dictatorial rule. Then ϕ is straightforward.*

Proof. Consider first the case where $P = A = \mathfrak{R}$.

The strategy of the proof is as follows: We consider three exhaustive and exclusive cases. The first case is when agent j's true preference $\bar{p}_j$ is such that individual j is a dictator when telling the truth, i.e. $g(\bar{p}_j, p_{-j}) = \bar{b}_j$, the bliss point of $\bar{p}_j$. The second case is when there is no preference that j can announce such that he or she is a dictator, i.e. $\forall p_i, g(p_i, p_{-j}) \neq \bar{b}_j$. Finally, the third case is when $\bar{p}_j$ is such that j is not a dictator when telling the truth, but can become a dictator by misrepresentation, i.e. there exists a $p_j' \neq \bar{p}_j$ such that $g(p_j', p_{-j}) = b_j'$ where b_j' is the bliss point of p_j'. In each of these three cases, we show that truthful revelation is a dominant strategy.

Assume that individuals' true preferences are given by the profile $(\bar{p}_1, \ldots, \bar{p}_k)$. We wish to prove that $\bar{p}_j$ is a dominant strategy for the j-th. individual.

Define $D_j \subset P^k$ to be the region where j is dictatorial. For any $k - 1$ tuple of strategies of agents other than j, denoted p_{-j}, there are three mutually exclusive and exhaustive cases:

(a) $(\bar{p}_j, p_{-j}) \in D_j$, i.e., $\phi(\bar{p}_j, p_{-j}) = \bar{b}_j$, where $\bar{b}_j$ is the bliss point of $\bar{p}_j$. Agent j is a dictator when telling the truth.

(b) $g(p_j, p_{-j}) \neq b_j$ for any $p_j \in P$, i.e. $(p_j, p_{-j}) \notin D_j$ for any $p_j \in P$. Agent j is never dictatorial.

(c) $g(\bar{p}_j, p_{-j}) \neq \bar{b}_j$, but there exists some $p_j' \in P$ such that $g(p_j', p_j) \neq b_j'$ (i.e. $(\bar{p}_j, p_{-j}) \notin D_j$ but $\exists p_j' \in P$ s.t. $(p_j', p_{-j}) \in D_j$). Agent j is not dictatorial when telling the truth, but can become "dictatorial by misrepresentation."

In case (a) it is obvious that $\bar{p}_j$ (i.e. the truth) is a dominant strategy for j. These cases are illustrated in Fig. 7. In case (b) let

$$\Psi(p_{-j}) = \{p_j \in P: (p_j, p_{-j}) \in P^k\}$$

In the set $\Psi(p_{-j})$ only p_j varies: by assumption, g is not dictatorial with dictator j in this set. Hence g fails to be constant with respect to p_j only on a set of measure zero in $\Psi(p_{-j})$. By continuity and because we are in case (b) the set $\Psi(p_{-j})$ has only one connected component. This implies $g(., p_{-j})$ must be a constant on all of $\Psi(p_{-j})$, which implies that the true message $\bar{p}_j$ is as good a strategy as any in P for the j-th. individual.

In case (c), consider the set $\tilde{D}(p_{-j})$ of strategies in P for the j-th. individual

$$\tilde{D}(p_{-j}) = \{q_j \in P: (q_j, p_{-j}) \in D_j\}.$$

$\tilde{D}(p_{-j})$ is thus the set of strategies that make j dictatorial within $\Psi(p_{-j})$. Consider now the strategy $\tilde{p}_j$ in $\tilde{D}(p_{-j})$ which is nearest in terms of the distance $d(\ldots)$ in $\mathfrak{R}$ to the true preference $\bar{p}_j$ (see Fig. 7), and let

$$d_0 = d(\tilde{p}_j, \bar{p}_j) = \min_{q_j \in \tilde{D}(p_{-j})} (d(q_j, \bar{p}_j)), \tag{2}$$

Note that $d_0 \neq 0$ by the construction of case (c). Outside of $\tilde{D}(p_{-j})$, $g(., p_{-j})$ is constant on any connected subset of $\Psi(p_{-j})$, by continuity. Hence it is

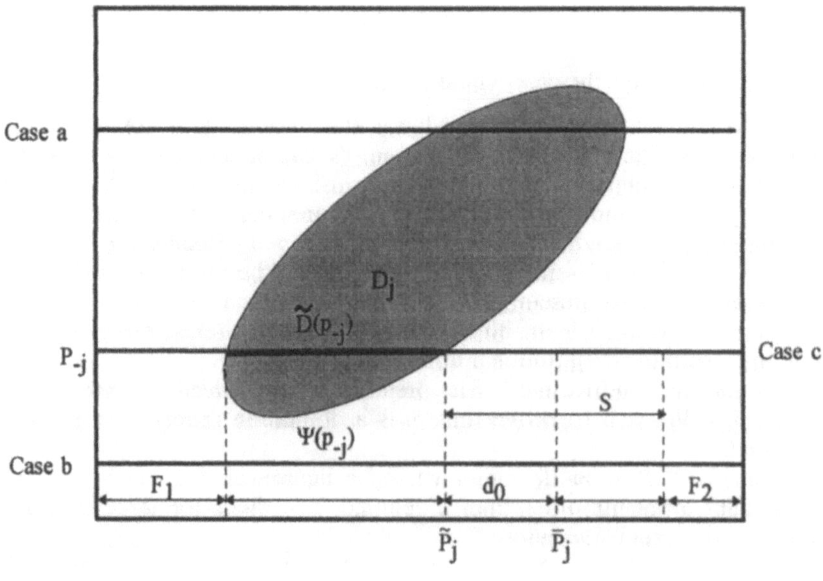

constant on the following connected set S, a set of outcomes not attainable by player j given the strategies p_{-j}:

$$S = \{q \in P: d(q, \bar{p}_j) \leq d_0\} \subset P,$$

so that by (2) for all q in S, $g(q, p_{-j}) = g(\tilde{p}_j, p_{-j})$ by continuity. In particular, $g(\bar{p}_j, p_{-j}) = g(\tilde{p}_j, p_{-j})$. Now, if $q_j \in \tilde{D}(p_{-j})$ is such that $d(\bar{p}_j, q_j) > d_0$, obviously on the set $\Psi(p_{-j})$, q_j is a strategy with a less desirable outcome for individual j than $\tilde{p}_j$, and therefore, also a less desirable outcome than the truth $\bar{p}_j$.

Assume without loss of generality that $\tilde{p}_j < \bar{p}_j$ and consider the set

$$F_1 = \{q \in P: q \leq \min\{p: p \in \tilde{D}(p_{-j})\}\}$$

Then $g(., p_{-j})$ is a constant map for all strategies in F_1. Since $d(F_1, \bar{p}_j) \geq d(\tilde{D}(p_{-j}), \bar{p}_j) \geq d_0$, and for $q_j \in F_1$, $g(q_j, p_j)$ is equal by continuity to $g(\tilde{q}_j, p_{-j})$ for some $\tilde{q}_j$ in $\tilde{D}(p_{-j})$, it follows that strategies in F_1 have a less desirable outcome than the truth $\bar{p}_j$. Now consider $F_2 = \{q \in P: q \geq \max\{q \in S\}\}$. This is a connected set on which g is constant. By continuity the outcome is equal to $g(\tilde{p}_j, p_{-j}) = g(\bar{p}_j, p_{-j})$. This completes the proof of straightforwardness when $P = C = I$.

Consider now $P = A = \mathfrak{R}^n$. Then, by definition since g is locally constant or dictatorial it is separable, i.e.

$$g(p_1, \ldots, p_k) = g_1(b_1^1, \ldots, b_k^1), \ldots, g_n(b_1^n, \ldots, b_k^n),$$

where b_i^j denotes the j-th. component of individual i's bliss point. Since the arguments given above apply to each $g_i: (\mathfrak{R}^k) \to \mathfrak{R}$, $i = 1, \ldots, n$, it follows that each g_i is straightforward, so that g is straightforward. This completes the proof of the proposition. ■

A.1.2. Proof of theorem 2

We now prove the main result of the paper, the equivalence of straightforwardness to being locally constant or dictatorial. The sufficiency of being LCD was of course established in Lemma 5, so that what remains is the necessity of being LCD.

Theorem 2. *A map g is straightforward if and only if it is locally constant or dictatorial* (*LCD*).

The strategy of the proof is as follows.

1. First we prove that being LCD is necessary for straightforwardness when the choice space is one dimensional, so that a game is a map from $\mathfrak{R}^k$ to $\mathfrak{R}$. In this case all metrics on the choice space agree and so preferences are characterized fully by their bliss points.
2. We then extend the result to higher dimensional cases. First we do this just for the case in which agent's strategies consist solely of announcing bliss points, and show that in this case any straightforward rule must be separable in the sense that the i-th. coordinate of the outcome depends only on the i-th. coordinates of the agents' strategies. In this case each coordinate function is a map from $\mathfrak{R}^k$ to $\mathfrak{R}$ and the results of the first case can be applied.
3. Next we analyze the case in which agents' strategies involve announcing the metrics of preferences as well as their bliss points. We show that in

this case the outcome of any straightforward rule must be unaffected by the metric announced, and so this case reduces to the previous one.

Step 1. Case $n = 1$, strategies are bliss points only.

Note that in the one dimensional case $P = \mathfrak{R}$, because preferences are statements of bliss points only, since all (non trivial) distances in $\mathfrak{R}$ are equivalent to the Euclidean distance. Assume that $g: S^k \to A$ is straightforward. By Theorem 1, g is continuous.

Let (m_i, m_{-i}) be a profile in $\mathring{P}^k$. Consider first the case where m_i is in the interior of the manipulation set $\mathring{M}_{m_{-i}}$. Then it follows by straightforwardness that

$$g(m_i, m_{-i}) = m_i. \tag{3}$$

Since g is continuous, if m^1_{-i} is a small variation of m_{-i}, m_i is also in $\mathring{M}_{m^1_{-i}}$, so that

$$g(m_i, m^1_{-i}) = m_i \tag{4}$$

for all m^1_{-i} in a neighborhood $U_{m_{-i}}$ of m_{-i} in S^{k-1}.

Similarly, continuity of g implies that if m^1_i is a small variation of m_i, m^1_i is in $\mathring{M}_{m^1_{-i}}$, so that (3) and (4) are also satisfied in a neighborhood of m_i.

We have therefore proven that for any profile $(m_i, m_{-i}) \in S^k$, if $m_i \in \mathring{M}_{m_{-i}}$, then g is dictatorial with dictator i in a neighborhood $W(m_i, m_{-i})$ of (m_i, m_{-i}) in S^k.

Consider now the case of a profile $(m_i, m_{-i}) \in S^k$ where $m_i \notin M_{m_{-i}}$ for all $i = 1, \ldots, k$.

In that case for any i

$$g(m_i, m_{-i}) = m \neq m_i. \tag{5}$$

Furthermore, by straightforwardness m is the best that the i-th. player can obtain, so that $m \in M_{m_{-i}}$ minimizes the distance between m_i and $M_{m_{-i}}$ in $A = \mathfrak{R}$. It follows that for m^1_i a small variation of m_i,

$$g(m^1_i, m_{-i}) = m, \tag{6}$$

since m will also minimize the distance in A between m^1_i and $M_{m_{-1}}$, see Fig. 8.

Therefore, g is a constant on a neighborhood V of m_i, within the pre-manipulation set $N_{m_{-i}}$, see Fig. 9.

Since the same argument is valid for all $i = 1, \ldots, k$, it follows that g is locally constant in a whole neighborhood $V(m_i, m_{-i}) \subset S^k$.

We have therefore proven that for any profile $(m_i, m_{-i}) \in S^k$, the map g is LCD whenever either $m_i \in \mathring{M}_{m_{-i}}$ for some i, or else $m_i \notin \mathring{M}_{m_{-i}}$, for all i. Note that since g is dictatorial with dictator i when $m_i \in \mathring{M}_{m_{-i}}$, then for any profile $(m_j, m_{-j}) \in P^k$, at most *one* message say m_j can be in the interior of the manipulation set determined by the others, $\mathring{M}_{m_{-j}}$. For any given profile $(m_i, m_{-i}) \in S^k$, there are therefore three exclusive and exhaustive cases:

(a) $m_j \in \mathring{M}_{m-j}$ for some m_j, a component of (m_i, m_{-i}) or

(b) $m_j \notin M_{m_{-j}}$ for all components m_j of (m_i, m_{-i}) or

(c) $m_j \in \partial M_{m_{-j}}$ for some component m_j of (m_i, m_{-i}).

As seen above, in case (a) the j-th. player is a dictator in a neighborhood of (m_i, m_{-i}), because the property $m_j \in \mathring{M}_{m_{-j}}$ is open in S^k. Therefore g is LCD at

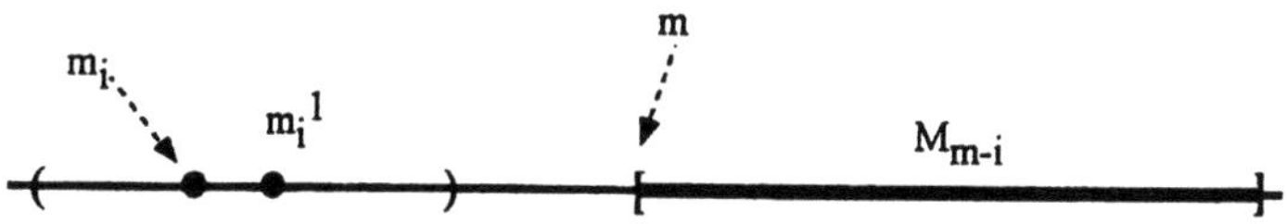

Fig. 8. Proof that g is straightforward $\Leftrightarrow$ it is LCD, when S is one dimensional. By straightforwardness the outcome m is locally independent of m_i

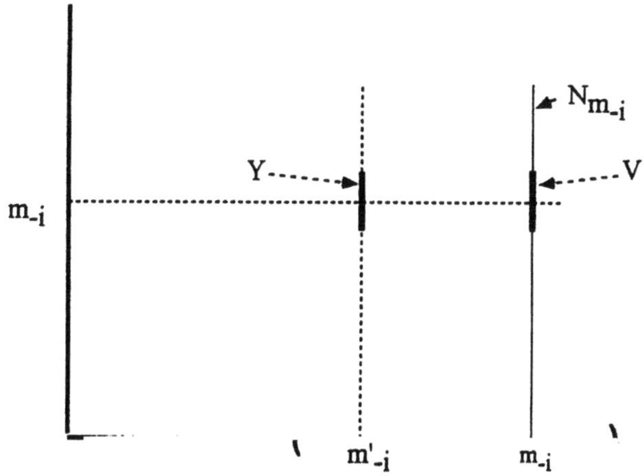

Fig. 9. On each premanipulation set g is locally constant in a neighborhood of m_i if $g(m_i, m_{-i}) \neq m$

such profiles. In case (b), g is locally constant in a neighborhood of (m_i, m_{-i}), as seen above. Therefore g is also LCD at such profiles.

In case (c) if there are at least two players, with messages m_j, m_k respectively and $m_j \in \partial M_{m_{-j}}$, $m_k \in \partial M_{m_{-k}}$, then m_j must equal m_k, since

$$m_j = g(m_j, m_{-j}) = g(m_k, m_{-k}) = m_k$$

by straightforwardness. Now, the space of profiles in S^k having at least two coordinates equal is a set of measure zero in S^k. Therefore case (c) is contained in a set of measure zero in S^k if $m_j \in \partial M_{m_{-j}}$ and $m_k \in \partial M_{m_{-k}}$ and $j \neq k$.

Now consider the last case in (c), where $m_j \in \partial M_{m_{-j}}$ for only one m_j in (m_i, m_{-i}). The manipulation sets $M_{m_{-j}}$ are closed intervals.[19] Therefore, the map assigning to each $k-1$ profile m_{-j} in S^{k-1} the strategy m_j in the boundary $\partial M_{m_{-j}}$ (a set consisting of two points in $\mathfrak{R}^1$) is the union of two continuous real valued maps on S^{k-1}. Since the graph of each of these maps is a set of measure zero in S^k, it follows that the set of profiles (m_k, m_{-i}) such that $m_j \in \partial M_{m_{-j}}$ for one $j \in \{1, \ldots, k\}$ has measure zero in S^k. Since this is true for each j, it follows that the set of profiles in case (c) have measure zero. This completes the proof that straightforward rules are LCD in the one-dimensional case. The converse has already been proven in Lemma 5: all LCD rules

[19] This is established in Lemma 4.

are straightforward. For the one-dimensional case the proof of the theorem is thus completed. We shall now reduce all more general cases to this case.

Step 2. $n > 1$, strategies are bliss points only.

We prove the theorem first for the special case *where the strategy space is just* $\mathfrak{R}^n$, *i.e.*, $S = \mathfrak{R}^n$ and the game form is $g:(\mathfrak{R}^n)^k \to A$, where A is a linear subset of $\mathfrak{R}^n$. This is the case in which agents' strategies are just the bliss points of preferences, and do not involve the statement of metrics.

The strategy of this proof is as follows: we show that if g is straightforward then g is a separable map, i.e., each coordinate g_i of the outcome

$$g(r_1, \dots, r_k) = (g_1(r_1, \dots, r_k), \dots, g_n(r_1, \dots, r_k))$$

depends only on the i-th. coordinates of the arguments r_k, i.e., on the vector $r_1^i, \dots, r_k^i$. So the i-th coordinate of the outcome depends only on the i-th coordinates of the bliss points announced. Once separability is established, the result that g straightforward implies that g is LCD follows immediately. This is because g is straightforward if and only if $g_i: \mathfrak{R}^k \to \mathfrak{R}$ is straightforward for all i, and for each $g_i:(\mathfrak{R})^k \to \mathfrak{R}$ the preceding proof (for the case $n = 1$) applies, so that g is straightforward if and only if it is LCD coordinate-wise.

Consider now $(r_1, \dots, r_k) \in (S)^k$. For any i, the manipulation set $M_{r_{-i}}$ is a convex set in A.[20]

Let $r_i \in \mathfrak{R}^n$ be i's bliss point, and let B_{r_i} be a ball centered on r_i, containing the point $r = g(r_1, \dots, r_k) \in A$. Let M_{r-i} be agent i's manipulation set: by straightforwardness r is the nearest point in M_{r-i} to r_i for any metric on $\mathfrak{R}^n$, corresponding to any preference with bliss point r_i (see Fig. 10). So for any ellipsoid E_{ri} centered on r_i and passing through r (B_{ri} is a particular case) there exists a hyperplane separating M_{r-i} from the interior of the ellipsoid. Hence the set of supports to M_{r-i} at r contains all possible tangents to ellipsoids centered on r_i and passing through r. This implies that M_{r-i} is contained in a cone based at r and generated by affine coordinate lines, as shown in Fig. 10a. Now it is easy to verify that M_{r-i} must equal the cone so generated rather than being strictly contained in it. This follows from straightforwardness and the fact that the outcome is by assumption independent of the metrics of agents' preferences. This is illustrated in figure 10b, where the nearest point in the manipulation set M_{r-i} to the bliss point r_i' depends on the metric around r_i'. This would not be true if M_{r-i} were generated by affine coordinate lines, as in the first panel.

It is clear that in this case, as shown in Fig. 10a, the k-th. coordinate of the outcome $r = g(r_1, \dots, r_k)$ depends *only* on the k-th. coordinates of the vectors r_i. Since this is true for all k, the separability of g is established. This completes the proof for the case $g:(\mathfrak{R}^n)^k \to A$, because each g_i must be LCD. In

[20] This can be easily seen from the arguments in Laffond [16], derived from those of Valentine [22], because $M_{r_{-i}}$ is, in this case, the manipulation set that would obtain if the i-th player's preference was required to have a Euclidean distance function (and any bliss point), which is the case studied by Laffond, see, e.g., his Theorem 1 and Lemmas 5 and 8. Note that the condition of anonymity of g is not required in these proofs, and that the proof applies for r_{-i} in $(R^n)^{k-1}$ as in our case.

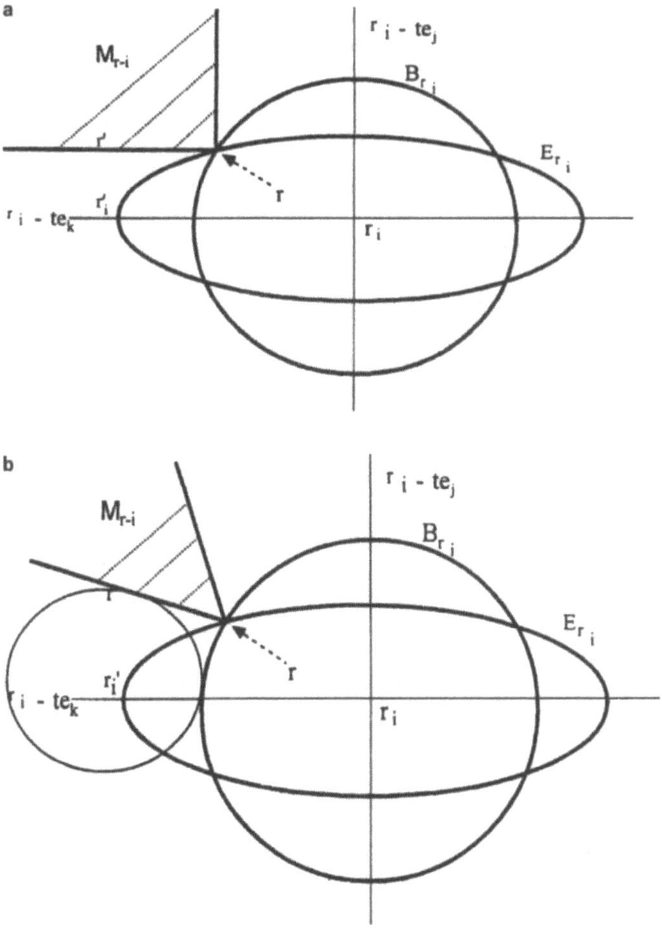

Fig. 10a, b. Proof that g is straightforward $\Leftrightarrow$ it is LCD. Case when $n > 1$. The manipulation set is the cone generated by the coordinate lines through r

particular, it shows that if $g:(\Re^n)^k \to A$ is straightforward, then g is also continuous, since each component g_i is.

Step 3. Strategies include metrics.

Finally, we consider the case where choices are in $\Re^n$, but strategies are preferences in P, so that g is a map defined on bliss points and metrics, $g:(P^n)^k \to A$, A a linear subset of $\Re^n$.

We break this step into two sub steps. In the first of these, we show that when restricted to preferences with bliss points in A, g only depends on these bliss points and *not* on the metrics announced by the players. This implies that g/A^k is actually a map from $A^k \subset (\Re^n)^k$ into $\Re^n$; therefore, we can apply the results of the previous case to show that g is separable, and thus the proof is completed for strategy profiles in A^k.

In the second sub step, we then show the proof is also valid for profiles in $(\Re^n)^k$ outside A^k, and the proof is completed.

Consider first the case in which the domain of the map g is preferences with bliss points in A. First we consider 2 players, and then we generalize this to any finite number of players. We proceed by induction on the number of coordinates in which the bliss points of their announced preferences differ.

The first case is when they differ only in one coordinate, i.e., p_1 and p_2 are two preferences with bliss points r_1 and r_2 and $r_1 = r_2 + \lambda e_1$, $\lambda > 0$, e_1 an element of the standard basis $\{e_i\}$ of $\mathfrak{R}^n$, which we call the first coordinate without loss of generality. Since g/A^k is onto, by lemma 1, g respects unanimity. Therefore player 1 could announce p_2 and get the outcome $g(p_2, p_2) = r_2$ and similarly player 2 could force the outcome r_1. So the outcome $g(p_2, p_2)$ must be preferred by 1 to r_2 and preferred by 2 to r_1. Hence it must be contained simultaneously in the ellipsoid E_{p_1} which is a contour of 1's preference and goes through r_2 and in the ellipsoid E_{p_2} which is a contour of 2's preferences and goes through r_1. This implies, in particular, by an extension of the arguments in the proof of Lemma 2, that the first coordinate of $g(p_1, p_2)$, denoted $g^1(p_1, p_2)$, is in the segment $[r_1^1, r_2^1]$, where r_1^1, and r_2^1 are the first coordinates of r_1 and r_2 respectively, see Fig. 11 below.

Thus, $g(p_1, p_2)$ must be contained in the singly shaded area in Fig. 11. Note that the outcome $g(p_1, p_2)$ must be contained in the line segment $[r_1, r_2]$, otherwise one can find a contradiction. To see this, consider a preference p_3 with a bliss point r_3 whose first coordinate is the same as that of $g(p_1, p_2)$ but which lies in the line segment $r_1 - r_2$. Now, $g(p_3, p_2)$ must be outside of E_{p_1} by straightforwardness, for otherwise an agent with true preference p_1 could announce p_3 and obtain an outcome preferable to $g(p_1, p_2)$. By the arguments of the previous paragraph, we must also have $g(p_3, p_2) \in E_{p_3}$ which is the ellipsoid centered on r_3 and through r_2, for otherwise a first agent with true preference p_3 could do better by announcing a preference p_2 and obtaining p_2 by respect of unanimity. This is true for any metrics and so for any ellipsoids E_{p_1} and E_{p_3} respectively centered on r_1 and r_3 and through r_2. This immediately establishes a contradiction, as is clear from Fig. 11. This contradiction cannot be established if $g(p_1, p_2) \in [r_1, r_2]$.

Now, if $g(p_1, p_2)$ were to vary within $[r_1, r_2]$ as the two players vary their metrics (keeping their bliss points r_1 and r_2 fixed) then obviously, the outcome could be manipulated by an appropriate choice of metric. Thus, when the bliss points of p_1 and p_2 differ in one coordinate only, g must be independent of the metric announced.

Furthermore, note that when r_1 and r_2 differ in one coordinate only, if $g(p_1, p_2) \in [r_1, r_2]$, the interior of the segment $[r_1, r_2]$, then $g(p_1, p_2)$ is a constant map for all p_1 and p_2 with bliss points r'_1 and r'_2 in the segment $[r_1, r_2]$: this follows from the characterization of straightforward games as LCD maps for the case $n = 1$, and the fact that as the outcome is neither r_1 nor r_2 so that neither player is dictatorial.

We now make the following *inductive hypothesis*:

Inductive assumption. (1) If p_1 and p_2 are two preferences whose bliss points r_1 and r_2 differ at most in m-1 coordinates, then the outcome $g(p_1, p_2)$ depends only on the bliss points r_1 and r_2 and, furthermore, (2) if for some coordinate j, $g^j(p_1, p_2)$ is in the interior of $[r_1^j, r_2^j]$, then g is constant for all (p_1, p_2) whose bliss points are in the box $B[r_1, \ldots, r_2]$ determined by r_1 and r_2, i.e.,

$$\{b \in \mathfrak{R}^n\colon \forall i, r_a^i \leq b^i \leq r_b^i, i = 1, \ldots, n \text{ where } \{a, b\} = \{1, 2\} \text{ or } \{2, 1\}\}.$$

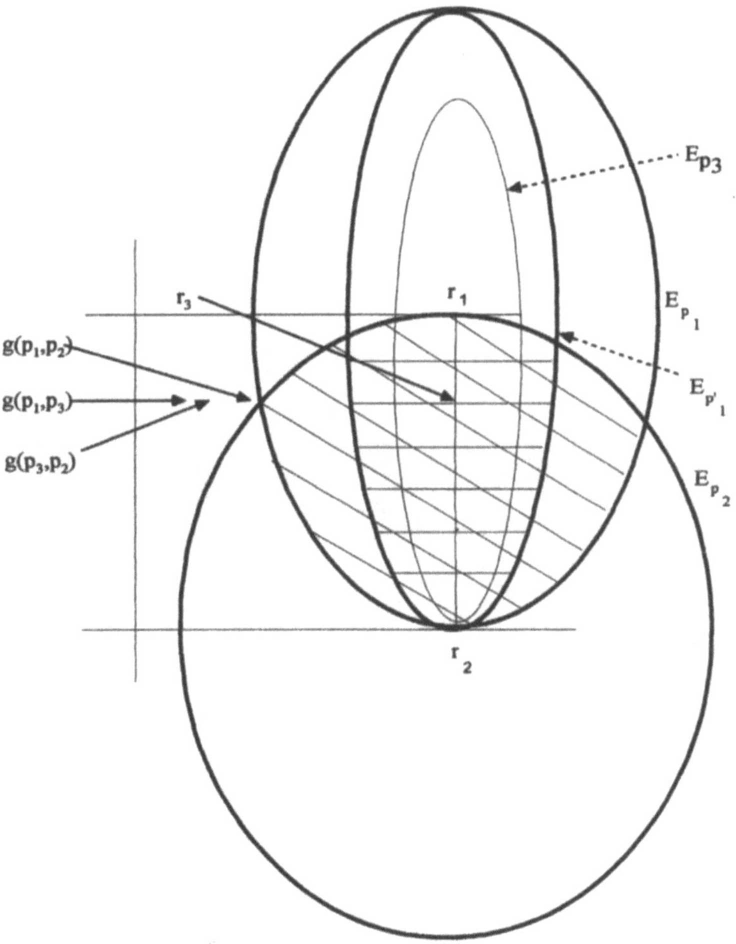

Fig. 11 Proof that g is straightforward $\Leftrightarrow$ it is LCD. $g(p_1, p_2)$ must lie in the shaded area

Now assume that the bliss points of p_1 and p_2, r_1 and r_2, differ in m coordinates. Two exclusive and exhaustive cases may arise: (a) $g(p_1, p_2)$ is contained in the box determined by r_1 and r_2, (2) this condition is not satisfied, so that we can assume without loss of generality that, for some coordinate j

$$g^j(p_1, p_2) > r_1^j > r_2^j,$$

see Fig. 12 below

In case (2), note that for all preferences p'_2 with bliss points s_2 in the half line $(r_2, g(p_1, p_2)]$ and same metric m_2 as p_2, we have

$$g(p_1, p_2) = g(p_1, p'_2);$$

this follows from the fact that as $g(p_1, p_2)$ is the nearest point to r_2 in the manipulation set of agent 2 according her metric, it is also the nearest point to

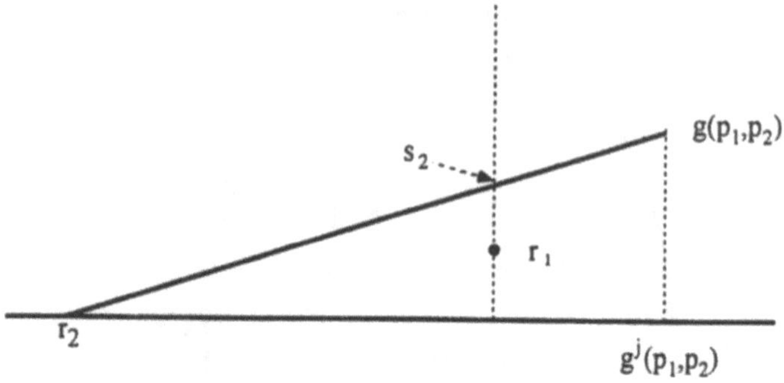

s_2 according to this metric (see Fig. 12). Now, choose p_2' so that it has a bliss point s_2 in $(r_2, g(p_1, p_2)]$ having the *same* j-th. coordinate as r_1, see Fig. 12. This is always possible because the half line $(r_2, g(p_1, p_2)]$ intersects the hyperplane $H^j = \{(x^1, \ldots, x^n): x^j = r_1^j\}$ in R^n.

Now, we assumed that the bliss points of p_1 and p_2 differ in m coordinates only, so the inductive hypothesis applies to p_1 and p_2' because p_2' and p_1 differ in m-1 coordinates only. Therefore,

1. $g(p_1, p_2) = g(p_1, p_2')$ by construction.
2. $g(p_1, p_2') = g(p_1, p_2'')$ where p_2'' has the same bliss point as p_2'' i.e., s_2, and any metric (using the inductive hypothesis).
3. $g(p_1, p_2'') = g(p_1, q_2)$ where q_2 is a preference which has a bliss point anywhere in the half line $(r_2, g(p_1, p_2)]$ and the same metric as p_2'', which may be any metric. This follows because $g(p_1, p_2'')$ is the nearest point to s_2 in the manipulation set of agent 2 according to her metric and so is the nearest point in the manipulation set to any point on the line $(r_2, g(p_1, p_2)]$ according to this metric.

It therefore follows that

$$g(p_1, p_2) = g(p_1, q_2)$$

for any preference q_2 with bliss point r_2, and arbitrary metric, which is what we wanted to prove – thus g is independent of the metric in this case also.

The only case left is (1). We can thus assume without loss of generality that for all j

$$r_1^j \leq g^j(p_1, p_2) \leq r_2^j,$$

Note that if for all p_1, p_2, one part of this inequality is an equality for all j, the result is automatically true because $g(p_1, p_2)$ is then in the boundary of the box determined by r_1 and r_2 so that its h-th coordinate depends only on the h-th coordinates of r_1 and r_2, i.e. it is separable. We can therefore assume a strict inequality for some j; without loss of generality, assume

$$r_1^j < g^j(p_1, p_2) < r_2^j.$$

The rest of the proof is simple: we show that we can alter p_1 and p_2 into $\bar{p}_1$ and $\bar{p}_2$ so that $\bar{p}_1, \bar{p}_2$ have bliss points which differ in $m - 1$ coordinates only and

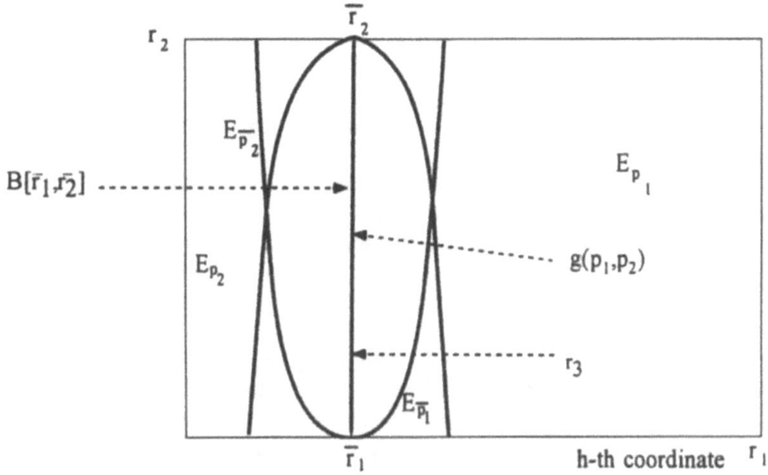

Fig. 13. g is constant for all p_1 and p_2 whose bliss points are in a box determined by r_1 and r_2

the condition

$$\bar{r}_1^j < g^j(p_1, \bar{p}_2) < \bar{r}_2^j$$

is still satisfied.

By (1) of the inductive hypothesis this implies $g(\bar{p}_1, \bar{p}_2)$ is independent of the metrics for all p_1, p_2 whose bliss points are in the $n-(m-1)$ dimensional box determined by $\bar{r}_1$ and $\bar{r}_2$. By (2) of the inductive hypothesis g is a constant for all preferences whose bliss points are in the $n-(m-1)$ dimensional box determined by $\bar{r}_1$ and $\bar{r}_2$. This will be shown to imply immediately that $g(p_1, p_2)$ is a constant in the box determined by r_1 and r_2; in particular the map g is independent of the metric in this case. Fig. 13 below illustrates the argument:

Consider $\bar{p}_1$ and $\bar{p}_2$ such that their bliss points $\bar{r}_1, \bar{r}_2$ have one more coordinate in common than do the bliss points r_1, r_2 of p_1 and p_2. Let this common coordinate be the h-th coordinate, and let this coordinate be the h-th coordinate of $g(p_1, p_2)$. Condition (1) of the inductive hypothesis is satisfied, so that g is independent of the metric if bliss points are in $B[\bar{r}_1, \bar{r}_2]$.

Since g is straightforward, the h-th coordinate of the outcome $g(\bar{p}_1, \bar{p}_2)$ is in the interior of the segment $(\bar{r}_1^h, \bar{r}_2^h)$ by the following arguments (which are an extension of those illustrated in Fig. 11 and used in the case of preferences whose bliss points differ in one dimension only at the start of step 3 of this proof, see Fig. 13 for an illustration). Suppose that $g(\overline{p_1}, \overline{p_2}) \notin B[\overline{r_1}, \overline{r_2}]$. Then consider the point in $B[\overline{r_1}, \overline{r_2}]$ nearest to $g(\overline{p_1}, \overline{p_2})$: call this point $\overline{r_3}$. Now we can generalize the one-dimensional argument at the start of step 3 of the proof which is illustrated in Fig. 13. Hence $g(\overline{p_1}, \overline{p_2}) \in$ *interior* $B[\overline{r_1}, \overline{r_2}]$, and of course the outcome is independent of the metrics announced.

Now we can use (2) of the inductive assumption to assert that g is also constant within the $n-(m-1)$-dimensional box determined by $\bar{r}_1$ and $\bar{r}_2$. In figure 13 this implies $g(p_1, p_2)$ is constant for p_1, p_2 whose bliss points are in $B[\overline{r_1}, \overline{r_2}]$, as in this case $g(p_1, p_2) \in B[\overline{r_1}, \overline{r_2}]$. A similar argument can be given to prove that $g(p_1, p_2)$ is constant along *any* coordinate segment j in which the inequality

$$r_1^j < g^j(p_1, p_2) < r_2^j$$

is satisfied and then that $g(p_1, p_2)$ is constant in the box determined by the coordinates of r_1 and r_2.

In summary: we have proven that if preferences have bliss points r_1, r_2 which differ in one coordinate only, then the outcome is in the interval defined by these bliss points and is independent of metrics: it is a constant for all bliss points in the interval $[r_1, r_2]$. We have then assumed as an inductive hypothesis that if preferences have bliss points r_1, r_2 which differ in at most $m-1$ coordinates, the outcome depends only on the bliss points and is constant whenever bliss points are in $B[r_1, r_2]$, provided that for some coordinate $g(p_1, p_2)$ is in the interior of $B[r_1, r_2]$. Given this assumption, we have proven that if bliss points differ in only m coordinates, the same properties hold. The completes the proof for 2 players. A straightforward induction procedure on the number of players shows that the result is also true for $k > 2$ players.

We therefore know that if g is straightforward, g is independent of the metric announced and depends only on bliss points. By the proof in step 2 of the first part of the case $n > 1$, this implies that g is separable on A^k, thus implying that g is locally constant or dictatorial on A^k.

Consider finally a set of strategies $(p_1, \ldots, p_k) \in (R^n)^k - A^k$. Assume first that a subset of preferences, say $p_1, \ldots, p_j, j < k$ does not belong to A. Then

$$g(p_1, \ldots, p_k) = g(\pi(p_1), \ldots, \pi(p_j), \ldots, p_k),$$

where $\pi(p_i)$ is the preference with same metric as p_i and bliss point in the intersection of ∂A and the half line $(r_i, g(p_1, \ldots, p_k)]$. Since, by construction $(\pi(p_1), \ldots, \pi(p_j), \ldots,) \in A^k$ then it follows by the first part of this proof that $g(\pi(p_1), \ldots, \pi(p_j), \ldots, p_k)$ does not depend on the metric of the preferences $\pi(p_1), \ldots, (p_j), \ldots, p_k$. Hence neither does $g(p_1, \ldots, p_k)$ depend on these metrics. Therefore the map g is independent of the metrics in this case also. So we have shown that if g is straightforward then g is separable, by step 2 of this proof.

Now, since for all $(p_1, \ldots, p_k) \in (R^n)^k$

$$g(p_1, \ldots, p_k) = g(\pi(p_1), \ldots, \pi(p_j), \ldots, p_k),$$

where $\pi(p_i) \in \partial A$ by construction, then, for all i, the manipulation set M_{P-i} corresponding to any profile $(p_1, \ldots, p_k)$ in $(R^n)^k$ is an affine subspace: this is because the first part of this proof for profiles in A^k now applies. The rests of the argument then follows: since for each i, $g(p_1, \ldots, p_{-i})$ minimizes the distance between r_i and M_{P-i} (where r_i is the bliss point of p_i), it follows that the map $g(p_1, \ldots, p_k) = g(r_1, \ldots, r_k)$ is separable in this one as well. The first part of the proof of the theorem (for $n = 1$) is now applied and it proves that g straightforward implies g is LCD. The converse follows from Lemma 5. This completes the proof of the characterization theorem. ■

A.2. Proofs of non-robustness

A topological space X is *second countable* if it has a countable base of neighborhoods for its topology.

A continous map $f: X \to Y$, X and Y topological spaces is called *open* if the image of any open set U in $X, f(U)$, is open in Y. Note that open maps have the property that if D is a dense subset of Y, then $f^{-1}(D)$ is dense in X.

A set is *residual* if it is the intersection of (at most) countably many open dense sets. The Baire Category theorem asserts that a residual subset of a complete metric space is dense.

Let $H = \{f: I^k \to \mathfrak{R}, f \text{ a bounded } C^{k+1} \text{ map}\}$. H is a linear space, with the addition rule $(f+g)(x) = f(x) + g(x)$. The C^{k+1} sup norm $\|.\|_{k+1}$ on H is defined by

$$\|f - g\|_{k+1} = \sup_{x \in I^k}\left(\sum_{j=0}^{k+1} \|D^j f(x) - D^j g(x)\|\right).$$

where $D^0(f)$ denotes f.

Endowed with the C^{k+1} norm, H is a Banach space, and in particular, a complete metric space.

Theorem 3. *The set of non-straightforward games on a bounded choice space is a residual set of the space of all continuous maps from $I^{n,k}$ to I^n, $C^0(I^{n,k}, I^n)$, and in particular is a dense set.*

Proof. The strategy of the proof is as follows. Let L be the set of continuous maps which are locally constant or dictatorial. We shall consider first the two simplest cases: when a rule g in L is *dictatorial*, and when it is *constant*. We prove that for any dictatorial rule g and dictator d, and any small $\varepsilon > 0$, there exists a map g_ε in the complement of L, $C(L) \subset C^0(I^{n,k}, I^n)$, with $\|g_\varepsilon - g\|_{\sup} < \varepsilon$. The proof will then be extended to include rules which are only locally dictatorial, or locally constant, thus implying that $C(L)$ is dense in $C^0(I^{n,k}, I^n)$. Finally, we shall prove that $C(L)$ is open.

Let $C(\nabla) = \{p \in I^k: p = (p_1, \dots, p_k), p_i \neq p_j \text{ for } i \neq j\}$.

Let $g: I^{n,k} \to I^n$ be a dictatorial map, with dictator d. For any $\varepsilon > 0$, $\varepsilon < 1/2$, let f be a C^1 diffeomorphism $f: I^n \to I^n$, such that

$$\sup_{b \in I} \|(f(b) - b)\| < \varepsilon,\ Df(b) \neq 0 \text{ for all } b \in I^n$$

$$\text{and} \tag{7}$$

$$f(b_d) \neq b_d \quad \text{for some } b_d \in I^n.$$

Consider now the composition map

$$g_\varepsilon = f \circ g: I^{n,k} \to I^n.$$

Then, by construction $\|g_\varepsilon - g\|_{\sup} < \varepsilon$. Now, since for all $p \in I^{n,k}$, $Df(p) \neq 0$ and $Dg(p) \neq 0$, it follows that $Dg_\varepsilon(p) \neq 0$ for all p. Therefore g_ε is nowhere locally constant.

Consider now $p \in g^{-1}(b_d) \subset I^{n,k}$, $p \in C(\nabla)$. Then $g_\varepsilon = f \circ g(p) = f(b_d) \neq b_d$ by construction. Therefore g_ε is not dictatorial because if it were dictatorial then $g(p_1, \dots, p_k) = p_d$ for some d, whatever p_d, and $g_\varepsilon \neq p_d$ for some $b_d \in I$.

Finally, g_ε is not locally dictatorial (with dictator other than d) at p, because by construction

$$\frac{\partial g_\varepsilon}{\partial p_j}(p) = Df(p) \cdot \frac{\partial g}{\partial p_j}$$

and $\partial g/\partial p_j = 0$ if $j \neq d$. Note that for all j, $\partial g/\partial p_j$ exists because g is dictatorial and, in particular, differentiable. Therefore g_ε is in $C(L)$. Since ε is arbitrarily chosen, any dictatorial rule g is a limit of rules in $C(L)$.

Now, consider any constant rule $g: I^{n,k} \to I^n$, $g(p_1, \ldots, p_k) = b_0 \in I^n$, b_0 a constant. By the Stone Weierstrass theorem, for any $\varepsilon > 0$ there exists a C^1 map g_ε such that

$$\|g_\varepsilon - g\|_{\sup} < \varepsilon \tag{8}$$

because $I^{n,k}$ is compact. We can obviously require, furthermore, that $Dg_\varepsilon(p) \neq 0$ for $p \in C(\nabla)$. Therefore g_ε is not locally constant at p.

Note that such g_ε cannot be locally dictatorial at p either because it is in an ε-neighborhood of the constant map g, and ε is arbitrarily chosen: any locally dictatorial map on a set $U \subset C(\nabla)$ will be at least at a positive distance ε_0 from the constant rule g, ε_0 a constant depending on the set U and on b_0.

Since ε is arbitrarily chosen, any constant rule g is a limit of rules in $C(L)$.

Consider now an arbitrary straightforward g in L, and let $p \in C(\nabla)$. Then there exists a neighborhood U of p such that either $g/_U$ is dictatorial, or $g/_U$ is a constant map.

The argument given above for constant and dictatorial maps, when restricted to $U \subset P^k$, prove that $g/_U$ can be arbitrarily approximated by a rule $g_\varepsilon(U)$ defined on U, and such that $g_\varepsilon(U)$ is neither locally constant nor locally dictatorial on U. A standard argument using partitions of unity (see, e.g. Guillemin and Pollak [14, p. 52]) can then be used to prove the existence of a continuous map $g_\varepsilon: I^{n,k} \to I^n$ such that $g_\varepsilon/_U = g_\varepsilon(U)$ and $\|g_\varepsilon - g\|_{\sup} < \varepsilon$.

Since $g_\varepsilon(U)$ is not locally constant nor locally dictatorial at p, g_ε is not either. Thus g_ε is a function in $C(L)$ within an ε-neighborhood of g. Therefore $C(L)$ is dense in $C^0(I^{n,k}, I^n)$.

Next we prove that it is open. Consider now $g \in C(L)$. Let $p \in C(\nabla)$ be such that g is neither locally constant, nor locally dictatorial at p.

The fact that g is not locally dictatorial at profile p implies that in any neighborhood U_p of p, there exists for each $j = 1, \ldots, k$ a profile p^j and a number $\varepsilon^j > 0$ such that

$$|g(p^j) - g_j(p^j)| = \varepsilon^j,$$

where $b_j(p^j)$ is the bliss point of the j-th. preference in the profile p^j. Now for any $\rho \in C^0(I^{n,k}, I^n)$,

$$|\rho(p^j) - b_j(p^j)| \geq \varepsilon^j - \|\rho - g\|_{\sup}.$$

From this inequality it follows that no rule within an $\varepsilon^j/2$ neighborhood of g in $C^0(I^{n,k}, I^n)$ can be locally dictatorial with dictator j at p. This is because any such rule will satisfy

$$|\rho(p^j) - b_j(p^j)| \geq \varepsilon^j/2$$

at any profile p^j where g satisfies

$$|\phi(p^j) - b_j(p^j)| = \varepsilon^j,$$

and every neighborhood of p will contain such points. If we now set $\varepsilon = \min_{j=1,\ldots,k} \varepsilon^j$, then no rule ρ within an $\varepsilon/2$ neighborhood of g in $C^0(I^{n,k}, I^n)$ is locally dictatorial at p.

As above, let $p \in C(\nabla)$, $g \in C(L)$ not locally constant at p, and let $q \in U_p$ be such that

$$|g(p) - g(q)| > \varepsilon^1$$

for some $\varepsilon^1 > 0$; such a q exists because g is not locally constant. Then

$$|\rho(p) - \rho(q)| > \varepsilon^1 - 2\|\rho - g\|_{\text{sup}}.$$

Therefore for $\delta = \frac{1}{2}\min(\varepsilon, \varepsilon^1)$, any rule ρ within a δ neighborhood of g in the sup topology is neither locally constant nor locally dictatorial at p. Therefore $C(L)$ is open, completing the proof. ■

A.2.1. Results on Nash equilibria

Theorem 4. *Let ϕ be a continuous social choice rule, $\phi: P^k \to A$, where $P = A = I$, the unit interval in $\Re$, and $k \geq 2$. Let $M = I$ be the message space consisting of statements on bliss points of individual preferences. If the rule ϕ is Nash implementable by a regular game $g: M^k \to A$, then ϕ is locally constant or dictatorial (LCD).*

Proof. Consider the reaction set R_{p_i} corresponding to individual i with preference $p_i \in P$, i.e., the set of message-profiles

$$R_{p_i} = \{(m_i(m_{-i}), m_{-i}) \in M^k \colon (m_i(m_{-i}), m_{-i}) \in \arg\max_{m_i \in M} (p_i(m_i, m_{-i}))\}.$$

This is the set of vectors $m \in M^k$ such that agent i's message is his or her optimal response to the messages of the other agents, i.e., to m_{-i}. It is a generalization of the concept of reaction function. Figure 14 illustrates this for two agents: the level sets of a game $g: \Re^2 \to \Re^1$ are shown, and agent 1's reaction set is the locus of points of tangency between these curves and horizontal lines corresponding to strategy choices by agent 2.

The strategy of the proof is as follows.

1. In step one we show that if preferences are such that a Nash equilibrium message profile in M^k is one at which no agent attains her bliss point, i.e., $b_i \neq g(\overline{m}) \forall i$, then the social choice rule is locally constant in a neighborhood of these preferences. This follows from two facts: that $b_i \neq g(\overline{m}) \forall i$ is an open condition, and that the profile of agent's messages at a Nash equilibrium satisfies simultaneous optimality conditions as each is a best response to the others. We show that these same optimality conditions continue to characterize agents' best responses for small changes in preferences, and that in this case the outcome of the social choice rule must be locally independent of the agents' preferences.
2. In step two we consider the case in which a Nash equilibrium gives as an outcome the bliss point of one agent, and show that in this case the social choice rule is locality dictatorial with that agents as dictator.

Step 1. No agent attains her bliss point at a Nash equilibrium

Now consider the set $T_{p_i} = R_{p_i} - g^{-1}(b_i)$, which is agent i's reaction set minus the preimage of her bliss point.[21] Let Π_{-i} be the projection of a vector

[21] Obviously, if $b_i \notin g(M^k)$ then $T_{p_i} = R_{p_i}$.

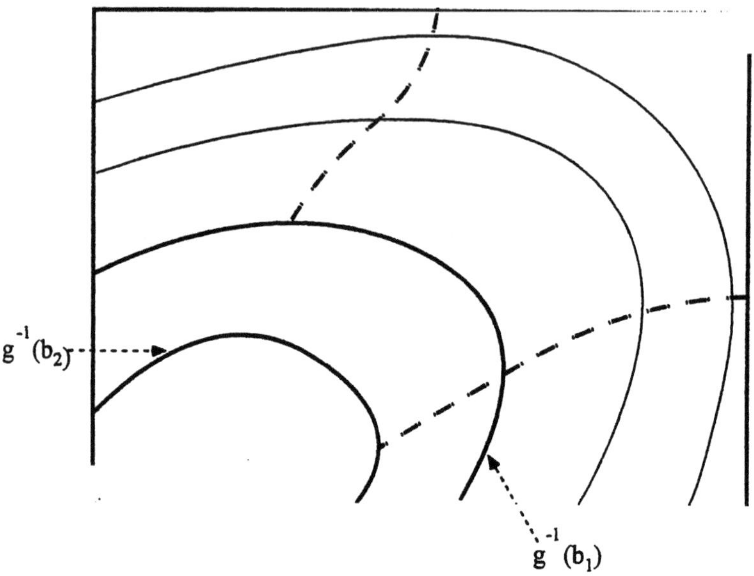

Fig. 14. The sets $g^{-1}(b_i)$, $i = 1, 2$

in M^k onto all coordinates other than the i-th. Note that for any set of strategies of players other than i, $m_{-i} \in M^{k-1}$, either this is in the projection onto coordinates other than the i-th of the preimage of i's bliss point, i.e., $m_{-i} \in \Pi_{-i}(g^{-1}(b_i))$ or it is not, i.e., $m_{-i} \in \Pi_{-i}(T_{p_i})$. In the first case, agent i's best response is obvious: it is m_i. Consider next the second case and let $m_{-i} \in \Pi_{-i}(T_{p_i})$. Since by definition of T_{p_i}, the i-th. individual is unable to obtain b_i as an outcome of the game in response to m_{-i}, i's preference-maximizing response is a message that minimizes the difference

$$g(m_i, m_{-i}) - b_i,$$

by definition of the preferences p_i in P.

For all $m_{-i} \in M^{k-1}$, the manipulation set $\mathscr{M}_{m-1}$[22] is a connected set in $\Re$: this is an application of the mean value theorem to the continuous real valued function of one variable $g(., m_{-i}): I \to \Re$. Therefore, since $(m_i, m_{-i}) \in g^{-1}(b_i)$, the problem

$$\min_{m_i \in M} (g(m_i, m_{-i}) - b_i)$$

is necessarily equivalent to one of the following optimization problems:

a) $\max_{m_i \in M} g(m_i, m_{-i})$ if $\sup \mathscr{M}_{m_{-i}} < b_i$,

[22] We are using a script letter, $\mathscr{M}$, to denote the manipulation set in this section to avoid confusion with the message space M.

or

b) $\min_{m_i \in M} g(m_i, m_{-i})$ if $b_i < \inf \mathcal{M}_{m_{-i}}$.

Since by assumption g is C^{k+1}, any optimal $m \in T_{p_i}$ satisfies first order orthogonality condition on its gradient Dg. If (m_i, m_{-i}) is in the interior of M^k, $\overset{\circ}{M}{}^k$, then

$$Dg \perp N_{m-i}, \tag{9}$$

where $N_{m-i} = \{(m_i, m_{-i}) \in S^k: m_i \in M\}$ is the premanipulation set of agent i given the strategies m_{-i} of other agents. It is the intersection of one coordinate axis in $\mathfrak{R}^k$ with M^k. Therefore, $Dg \perp N_{m-i}$ implies $Dg_i = 0$ if $m = (m, m_i) \in \overset{\circ}{M}{}^k$. If instead, m is in the boundary of M^k, ∂M^k, then m belongs to a face F_β, i.e. a subset of M^k characterized by having all coordinates except for those in the set $\beta \subset \{1, \ldots, k\}$ constant, and equal either to zero or to one. In this case, the orthogonality condition for optimality is

$$\Pi_\beta(Dg) \perp N_{m-i}, \tag{10}$$

where Π_β denotes the projection map from $\mathfrak{R}^k$ into $\mathfrak{R}^\beta$, the Euclidean space with coordinates in β. By definition, this latter orthogonality condition implies that the i-th. coordinate of $\Pi_\beta(Dg)$ must vanish, i.e. that $Dg_i = 0$ if $i \in \beta$. Note that in addition to (9) and (10), the solutions to problem (a) and (b) must satisfy second order conditions, and must be global. Let $g(., m_{-i}): I \to I$, and $\bar{m} = (\bar{m}_i, \bar{m}_{-i})$ be in T_{p_i}. Then in the case of (a) $\partial^2 g(\bar{m})/\partial m_i^2 \leq 0$ for all i. Now taking $\eta = \{i\}$, the regularity condition (1) implies $g_i \pitchfork \Delta_i$. Note that $(\bar{m}, g(\bar{m})) \in g_i^{-1}(\Delta_i)$ since $g_i(\bar{m}, g(\bar{m})) = (Dg_i(\bar{m}) + g(\bar{m}),\ g(\overline{\overline{m}})) = (g(\bar{m}),\ g(\bar{m}))$. Therefore, by (1) $g_i(\bar{m}, f(\bar{m})) \pitchfork \Delta_i$, i.e., $D(Dg_i(\bar{m}) + g(\bar{m})) \neq Dg(\bar{m})$, implying $\partial^2 g(\bar{m})/\partial m_i^2 \neq 0$. Therefore, in the case of (a) if $\bar{m} \in T_{p_i}$

$$\frac{\partial^2 g(\bar{m})}{\partial m_i^2} < 0. \tag{11}$$

Similarly, in the case of (b)

$$\frac{\partial^2 g(\bar{m})}{\partial m_i^2} > 0 \text{ if } \bar{m} \in T_{p_i}. \tag{12}$$

Consider now a profile $(\bar{p}_i, \ldots, \bar{p}_k)$ in P^k and let $\phi(\bar{p}_i, \ldots, \bar{p}_k) = \bar{c}$ in A. Since g Nash-implements ϕ by assumption, there exists at least one message profile denoted $\bar{m} = \bar{m}(\bar{p}_i, \ldots, \bar{p}_k) \in M^k$, which is a Nash equilibrium of the game form g with preferences over outcomes $(\bar{p}_i, \ldots, \bar{p}_k)$, satisfying

$$g(\bar{m}) = \phi(\bar{p}_i, \ldots, \bar{p}_k).$$

Now, by definition, $\bar{m} \in \cap_{i=1}^k R_{\bar{p}_i}$. Recall that, for each i, $R_{\bar{p}_i} = T_{p_i} \cup g^{-1}(\bar{b}_i)$, where $\bar{b}_i = \bar{b}_i(\bar{p}_i)$ is the bliss point of $\bar{p}_i$.

We shall consider first the case in which $\overline{m} \notin g^{-1}(\bar{b}_i)$, and show that $\bar{m}$ is a Nash equilibrium for any preference profile p in some neighborhood V of $\bar{p}$ in P^k. Since g Nash implements ϕ, this will imply that for all $p = (\bar{p}_i, \ldots, \bar{p}_k)$ in V,

$$\phi(\bar{p}_i, \ldots, \bar{p}_k) = g(\bar{m}) = \bar{c}$$

i.e. ϕ is a constant on V.

Observe that the orthogonality conditions (9) and (10) (valid for M^k and ∂M^k respectively) are only dependent on the gradient of the game form Dg at $\bar{m}$, and not on the chosen profile $\bar{p}$. With respect to the second order conditions (11) and (12) (associated with problems (a) and (b) respectively) these will be satisfied in some neighborhood w of $\bar{p}$ in P^k when they are satisfied at $\bar{p}$, since they are open conditions: small variations of the preference profiles $\bar{p}$ in W are associated with small variations of the corresponding bliss points $(\bar{b}_1, \ldots, \bar{b}_i)$, so that if for some j and $m_{\hat{j}}$, $\sup(F(m_{\hat{j}})) < \bar{b}_j$, then $\sup(F(m_{\hat{j}})) < b_j)$, for $b_j = b_j(p_j)$, and p_j in a ε-neighborhood N_ε of $\bar{p}_j$, and if for some i and $m_{\hat{i}}$ $\inf(F(m_{\hat{i}})) > \bar{b}_i$, then $\inf(F(m_{\hat{i}})) > b_i$ for $b_i = b_i(p_i)$, p_i in an ε-neighborhood N_ε of $\bar{p}_i$. Note that the ε's of N_ε can be chosen uniformly (for all $m_{\hat{j}}$ in M^{k-1}) because of the compactness of M^{k-1}.

Hence $\bar{m}$ satisfies both first and second order conditions for all profiles p in a neighborhood W of $\bar{p}$. In addition, the components of $\bar{m}$ will be globally optimal responses for all profiles near enough to p.[23] It follows that $\bar{m}$ is a Nash equilibrium of the game form g for all profiles of preferences p in some neighborhood V of $\bar{p}$, $V \subset W$. Thus ϕ is locally constant at $\bar{p}$ in this case.[24] This completes step one of the proof.

Step 2. i's bliss point is a Nash equilibrium.

Consider now the case in which the Nash-equilibrium set of messages $\bar{m}$ associated with a profile $\bar{p} = (\bar{p}_1, \ldots, \bar{p}_k)$ is in $g^{-1}(\bar{b}_d)$, for some $d = 1, \ldots, k$. We shall show that in this case the social choice rule ϕ is locally dictatorial. Note that if $\bar{m} \in g^{-1}(\bar{b}_i) \cap g^{-1}(\bar{b}_j)$, then $\bar{b}_i = \bar{b}_j$, since the hypersurfaces of a function do not intersect.

We now show that if for all i, $j = 1, \ldots, k$, $\bar{b}_i \neq \bar{b}_j$, then there exists a neighborhood U of $\bar{p}$ in P^k such that the Nash equilibria corresponding to any profile p in $U(\bar{p})$, $\bar{m}$, are also in $g^{-1}(b_d)$, where $b_d = b_d(p_d)$ is the bliss point of the d-th. preference p_d in p. Since g Nash-implements ϕ, this implies that $\forall p$ in $U(\bar{p})$, $\phi(\bar{p}_1, \ldots, \bar{p}_k) = g(\bar{m}(p)) = b_d$, i.e. ϕ is locally dictatorial at $\bar{p}$, with dictator d. This would complete the proof that ϕ is LCD when it is Nash-implementable by a regular game.

Since $\bar{m}(\bar{p}) \in g^{-1}(\bar{b}_d)$ by assumption, if $\bar{m} \in \mathring{M}^k$ it follows that for $\eta = \{1, \ldots, d-1, \ldots, k\}$ (the set of integers from 1 to k with d deleted) the couple $(\bar{m}, \bar{b}_d) \in g_\eta^{-1}(\Delta_\eta)$ i.e.,

$$(Dg_{\eta_1}(\bar{m}) + g(\bar{m}), \ldots, Dg_{\eta_{k-1}}(m) + g(\bar{m}), g(\bar{m})) = (\bar{b}_d, \ldots, \bar{b}_d) \in \Delta_\eta,$$

since $g(\bar{m}) = \bar{b}_d$, and for all agents $j \neq d$ messages will be chosen to satisfy orthogonality conditions. By the regularity assumption (1) $g_\eta \pitchfork \Delta_\eta$, and $g_\mu \pitchfork \Delta_\mu$ for any $\mu \subset \eta$, implying that the map ∂g_η, the restriction of g_η on $\partial(M^k \times I)$, satisfies $g_\mu \pitchfork \Delta_\mu$.

---

[23] Suppose $\sup(F(\bar{m}_{\hat{j}})) < \bar{b}_j$. Then as noted above $\sup(F(\bar{m}_{\hat{j}})) < b_j$ for $b_j(p_j)$, p_j in N_ε of $\bar{p}_j$. Hence if $\bar{m}_j$ solves the problem $\max_{m_j \in M} g(m_j, \bar{m}_{\hat{j}})$ globally, then it is the globally optimal response for any $p_j \in \bar{p}_j$.

[24] One can actually show that if $\bar{m} \in \cap_{i=1,\ldots,k} T_{\bar{p}_i} \forall \bar{p}_i \in p$, then $\exists$ a neighborhood N of $\bar{p}$ such that $m \in \cap_{i=1,\ldots,k} T_{p_i} \forall p_i \in p$ in $N(\bar{p})$. Since $\bar{m} \in \cap_{i=1,\ldots,k} T_{\bar{p}_i}$, then by definition $g(\bar{m}) \neq \bar{b}_i \forall \bar{b}_i$ corresponding to the profile $\bar{p}_i$, i.e., $\bar{m} \notin \cup_{j=1,\ldots,k} g^{-1}(b_j)$ for b_j sufficiently close to $\bar{b}_j$. Thus $\bar{m} \in R_{p_i} - g^{-1}(b_i) \forall p_i \in p \in N(\bar{p})$, thus implying by definition that for all $p \in N$, $\bar{m} \in \cap_{i=1,\ldots,k} T_{p_i}$, $p_i \in p$.

Now, $g_\eta : M^k \times I \to \Re^k$, and Δ_η is a one dimensional submanifold of $\Re^k$. Therefore, by the transversality theorem [14, 1974, p. 60], $g_\eta^{-1}(\Delta_\eta)$ is a one dimensional submanifold of $M^k \times I$ (possibly with boundaries and corners). Therefore, there exists a neighborhood U of $(\bar{m}, \bar{b}_d)$ in $M^k \times I$ and a C^1 curve $b_d \to (m(b_d), b_d)$, for all $b_d \in \Pi_{k+1}(U)$, the projection of U onto its $k+1$-th. coordinate, such that $(m(b_d), b_d)$ is contained in $g_\eta^{-1}(\Delta_\eta)$, i.e.

$$g(m(b_d)) = b_d \text{ and } Dg_{\eta_1}(m(b_d)) = \dots = Dg_{\eta_{k-1}}(m(b_d)) = 0.$$

We shall now show that U can be chosen sufficiently small that the C^1 curve $m(b_d)$ in M^k consists of Nash equilibria corresponding to preference profiles $p = (p_1, \dots, p_k)$ in some neighborhood of $\bar{p}$ in P^k.

We know that for all j in η, $Dg_j(m(b_d)) = 0$ because $(m(b_d), b_d) \in g_\eta^{-1}(\Delta_\eta)$ by construction. Therefore all message profiles in the curve $m(b_d) \subset \Pi_{-(k+1)}(U) \subset M^k$ satisfy the first order conditions (1) (which are independent of the preference profiles). Recall that $\Pi_{-(k+1)}$ is the projection map on all coordinates but $k+1$.

Consider now a profile $\tilde{p}$ of the form $(\bar{p}_1, \dots, \tilde{p}_d, \dots, \bar{p}_k)$ where all but the d-th. preference are as in the profile $\bar{p}$, and such that the bliss point b_d corresponding to $\tilde{p}_d$ is in $\Pi_{k+1}(U)$, the projection of U onto its $k+1$-th. coordinate. Then when U is sufficiently small, $m(\tilde{b}_d)$ is a Nash equilibrium for $(\bar{p}_1, \dots, \tilde{p}_d, \dots, \bar{p}_k)$. To see this, note first that $g(m(\tilde{b}_d)) = \tilde{b}_d$, so that the d-th. individual strategy is indeed optimal. Next note that $m(\tilde{b}_d)$ satisfies the first order conditions (9), (10) as shown above, and the second order conditions (11) and (12) corresponding to problems (a) and (b) by the openness of these conditions. By continuity and an argument similar to step one and that in the last footnote above, U can be chosen small enough that $m_j(\tilde{b}_d)$ is globally optimal for $j \neq d$. This proves the point.

We have therefore shown that any message $m(\tilde{b}_d)$ in the curve $m(b_d) \subset \Pi_{-(k+1)}(U)$ is a Nash equilibrium for a preference profile $\tilde{p} = (\bar{p}_1, \dots, \tilde{p}_d, \dots, \bar{p}_k)$ in a neighborhood U of $\bar{p} \in p^k$, where $\tilde{b}_d$ is the bliss point of $\tilde{p}_d$. Furthermore, we have also shown that any message in the curve $m(b_d) \subset \Pi_{-(k+1)}(U)$ is in the set $T_{\bar{p}_j}$ for all j within the set of indices η, where $\bar{p}_j$ is the j-th. preference in the profile $(\bar{p}_1, \dots, \bar{p}_k)$.

This implies that to each preference profile of the form $p = (\bar{p}_1, \dots, \tilde{p}_d, \dots, \bar{p}_k)$ in a neighborhood U of $\bar{p}$, corresponds to a Nash equilibrium in the curve $m(\tilde{b}_d) \subset \Pi_{-(k+1)}(U)$, namely $m(\tilde{b}_d)$. Note that there could be Nash equilibria other than $m(\tilde{b}_d)$ associated to $\tilde{p}$. However, in order to know the value of ϕ at $\tilde{p}$ it suffices to know that $m(\tilde{b}_d)$ is a Nash equilibrium of $\tilde{p}$: as g Nash implements g,

$$\phi(\tilde{p}) = g(m(\tilde{b}_d)) = \tilde{b}_d \quad \text{for all } \tilde{p} \text{ in } N \text{ of the form } \tilde{p} = (\bar{p}_1, \dots, \tilde{p}_d, \dots, \bar{p}_k)$$

We now use the results of step 1 above.

Given that $m(\tilde{b}_d)$ is a Nash equilibrium for $\tilde{p}$, and that d is a dictator at $m(\tilde{b}_d)$, it follows that for all $j \neq d$, $m(\tilde{b}_d)$ must be in $T_{\bar{p}_j}$. But as we saw in step 1 this implies that there exists for all $j \neq d$ an ε-neighborhood N_ε of $\bar{p}_j$ such that $m(\tilde{b}_d)$ is in R_{p_j} for any p_j in N_ε. We have therefore proven that for all profiles $p = (p_1, \dots, p_d, \dots, p_k)$ in some neighborhood W of $\bar{p}$ in P^k there exists a Nash equilibrium in $\Pi_{-(k+1)}(U)$, namely $m(b_d)$, where b_d is the bliss point of p_d. It follows by construction of the curve $m(b_d)$ that for all p in W,

$\phi(p) = g(m(b_d)) = b_d$, so that ϕ is dictatorial in W, with dictator d. This completes the proof of step 2, when $\bar{m} \in \mathring{M}^k$. A similar proof applies for $\bar{m} \in \partial M^k$, e.g. $\bar{m} \in \mathring{F}_\eta \subset M^k$, where F_η is face in M^k with all coordinates but those in η constant, since condition (1) applies for all η. This completes the proof of the theorem. ■

References

1. Abraham R, Robbins J (1967) Transversal mappings and flows. WA Benjamin
2. Barbera S, Gul F, Stacchetti E (1992) Generalized median voter schemes and committees. Working paper, Universität Autónoma de Barcelona, Barcelona
3. Barberià S, Massó J, Neme A (1994) Voting under constraints. Working paper, Universitat Autónoma de Barcelona, Barcelona
4. Border KC, Jordan JS (1983) Straightforward election, unanimity and phantom voters. Social Science Working Paper 376, California Institute of Technology (Jan. 1981) Rev Econ Studies 50(1): 153–170
5. Chichilnisky G (1967) Manifolds of preferences and equilibria. PhD. Thesis, University of California, Berkeley
6. Chichilnisky G (1980) Social choice and the topology of spaces of preferences. Adv Math 37(2): 165–176
7. Chichilnisky G (1982) Social aggregation rules and continuity. Q J Econ 97(2): 337–352
8. Chichilnisky G (1993) On strategic control. Q J Econ 108: 285–290
9. Chichilnisky G, Heal GM (1983) A necessary and sufficient conditions for the resolution of the social choice paradox. J Econ Theory 31: 68–87
10. Chichilnisky G, Heal GM (1979) "Incentive compatibility and local simplicity" and "incentives to reveal preferences: the case of restrincted domains". Essex Economic Papers, University of Essex
11. Chichilnisky G, Heal GM (1979) A necessary and sufficient condition for straightforwardness. Essex Economic Papers, University of Essex
12. Chichilnisky G, Heal GM (1984) Patterns of power: bargaining and games in two-person situations. J Public Econ 23: 333–349
13. Guesnerie R, Laffont JJ (1981) The robustness of strategy-proof rules. Discussion Paper, Ecole des Hautes Etudes des Sciences Sociales, Paris
14. Guillemin V, Pollack A (1974) Differential topology. Prentice-Hall, Englewood Cliffs, NJ
15. Heal GM (1995) Incentive compatibility with infinite populations. Working Paper, Columbia Business School
16. Laffond G (1980) Revelations des preferences et utilities unimodales. Laboratoire d'Econometrie du CNAM, Paris Cedes
17. Laffont JJ, Maskin E (1980) A differential approach to dominant strategy mechanism. Econometrica 48(6): 1507–20
18. Moulin H (1980) Strategy-proofness and single peakedness. Public Choice
19. Rasmussen H (1997) Strategy-proofness of continuous aggregation maps. Soc Choice Welfare 14: 249–257
20. Saari D (1997) Informational geometry of social choice. Soc Choice Welfare 14: 211–232
21. Satterthweite M, Sonnenschein H (1981) Strategy-proof allocation mechanisms at differentiable points. Rev Econ Studies 48(4): 587–97
22. Valentine F (1964) Convex sets. McGraw Hill
23. van der Stel J (1993) Strategy-proof voting schemes on Euclidean spaces. Universitaire Pers Maastricht
24. Zarantonello E (1971) Contributions to non linear functional analysis. Academic Press, New York

Homotopy properties of Pareto aggregation rules

Gleb Koshevoy

Russian Academy of Sciences, Central Economics and Mathematics Institute
Krasikova 32, Moscow 117418, Russia

Received: 30 December 1994/Accepted: 22 April 1996

Abstract. We establish that the Pareto property is inconsistent with non-dictatorship for social choice rules defined on two-dimensional choice spaces. In addition, we consider applications of this result in higher dimensions.

We also establish the existence of Pareto rules with infinite populations, and show that in this case there is a strong manipulator.

1. Introduction

Arrow's paradox shows that there is a contradiction between three apparently reasonable and desirable conditions on the aggregation of individual preferences: non-dictatorship, the Pareto condition, and the condition of independence of irrelevant alternatives. Chichilnisky (1982) proved that at the topological level, the Pareto condition and non-dictatorship are inconsistent, i.e. any continuous social choice rule that satisfies the Pareto condition and an additional condition of weak positive association can be continuously deformed into (is homotopic to) a dictatorial rule. In the case of two voters, weak positive association can be omitted. In general, without the positive association condition a continuous Pareto rule is homological to a dictatorial rule (this is the approach in Chichilnisky 1993). Homological equivalence is weaker than homotopy equivalence: it does not imply that the rule can be continuously deformed into a dictatorial rule. However, both results imply the existence of a strong manipulator: a rule has a strong manipulator if there exists a unique voter who can obtain (possibly by misrepresentation) any possible preference as an outcome, given any preferences announced by the other voters.[1]

The author thanks G.M. Heal for helpful suggestions.

[1] See also the discussion in Heal's introduction to this volume.

In Sect. 3 of this paper, we show that in the case of linear preferences on a two-dimensional choice space the homotopy equivalence results can be generalized. Theorem 3.1 shows that in this case the Pareto condition is topologically inconsistent with non-dictatorship (i.e., homotopic to a dictatorial rule) for any finite number of voters, without the condition of weak positive association. Applications of this result in higher dimensions are also considered.

We also consider the case of an infinite population of voters. We show (Theorem 4.1) that any Pareto aggregation rule continuous in the Tychonov topology is homologous to a dictatorial rule. Recall that in general rules which are unanimous, anonymous and continuous with respect to the Tychonov topology case fail to exist. This is shown in, for example, Efimov and Koshevoy (1994).

The main mathematical tools used in the paper are cohomological methods in obstruction theory. This allows us to describe the set of all homotopy classes of continuous maps in $K(\Pi, n)$ spaces: $K(\Pi, n)$ is the Eilenberg–McLane space whose n-th homotopy group is Π and all other are zero.

2. Notation and definitions

Consider the space of preferences in the Chichilnisky framework. Denote by X the choice set, which is assumed to be diffeomorphic to the closed unit ball B^n in $\mathfrak{R}^n$. A *preference* p is a C^1 vector field on X, $p: X \to \mathfrak{R}^n$, which is locally the gradient field of a utility function u on X. The vector $p(x)$ at the choice $x \in X$ is the normal to the tangent space of indifference surface of u at x. The vector field is normalized to be unit length ($\| p(x) \| = 1$ for all $x \in X$). The space of preferences on X denoted P is a subset of the Banach space $V(X)$ of all C^1 vector field on X. The space P endowed with the topology inherited from $V(X)$ is a complete space (see Chichilnisky 1976).

We consider l voters ($l \geq 2$, L denotes the set of voters). A *profile* $\mathbf{p} = (p_1, \dots, p_l)$ is an ordered l-tuple of preferences of the voters, so that space of profiles is P^L.

A *social aggregation* rule is a continuous map that maps each profile in P^L to a social preferences in P, $\phi: P^L \to P$.

A rule ϕ is *Pareto rule* if, when all voters prefer x to y, the social preference ranks x above y.

A rule ϕ *respects unanimity* if the superposition of the map ϕ with the diagonal map $\Delta(P) \subset P^L$ is the identity map on $P(\Delta(p) = (p, \dots, p)\ \forall p \in P)$. It is evident that a Pareto rule respects unanimity.

An aggregation rule ϕ_k is said to be *dictatorial* (with a dictator k) if ϕ_k is the projection onto the k-th coordinate,

$$\phi_k(p_1, \dots, p_l) = p_k, \quad \forall (p_1, \dots, p_l) \in P^L.$$

A rule is said to be satisfy the *weak positive association condition* (Chichilnisky 1982) if whenever for some $k \in \{1, \dots, l\}$ and a profile $\mathbf{p} \in P^L$ we have $\phi(p_1, \dots, p_l) = -p_k$, then $\phi(-p_k, \dots, -p_k, p_k, -p_k, \dots, -p_k) \neq p_k$. A rule is called a *W-Pareto* rule if it is a Pareto rule satisfying *wpac*.

We shall have occasion to consider rules which satisfy the following property, denoted $(*)$. Let $\phi(-p, \dots, -p, p, -p, \dots, -p) = p$ for some p at the i-th place: then for any profile $\mathbf{p}_{-i} = (p_1, \dots, p_{i-1}, p_{i+1}, \dots, p_l) \in \mathrm{P}^{L\setminus\{i\}}$ we have $\phi(p_1, \dots, p_{i-1}, p, p_{i+1}, \dots, p_l) \neq -p$.

Note that when rules under consideration are Pareto rules, the last property is equivalent to *wpac*. The property $(*)$ has the following interpretation: if the voters $1, \dots, i-1, i+1, \dots, l$ cannot attain the outcome $-p$ by announcing preferences $-p$ while p is the i-th participant's preference, then they can not receive $-p$ as the outcome by announcing any other preferences.

Let Y and Z be topological spaces, and f and g be continuous functions $f: Y \to Z$, $g: Y \to Z$. f is said to be *homotopic* to g when there exists a continuous map $\mathscr{H}: Y \times [0, 1] \to Z$, satisfying $\mathscr{H}(y, 0) = f(y)$ and $\mathscr{H}(y, 1) = g(y)$, for all y in Y. Denote $\mathscr{C}(Y, Z)$ the space of all continuous maps Y in Z. The set of the equivalence classes with respect to the homotopic relation on the set $\mathscr{C}(Y, Z)$ is denoted by $\pi(Y, Z)$ (for further references on algebraic topology, see, for example, Fomenko and Fucks 1989).

The following theorems have been proved by Chichilnisky (1982, 1993).

Theorem 2.1. *Let $\phi: P^L \to P$ be a continuous W-Pareto aggregation rule. Then ϕ is homotopic to a dictatorial rule.*

To state the next theorem, the following notions are needed. A participant i is said to be a *manipulator* if

$$\forall p \in P,\ \forall (p_1, \dots, p_{i-1}, p_{i+1}, \dots, p_l)\ \exists p_i \in P \text{ such that } \phi(p_1, \dots, p_l) = p.$$

A participant is said to be a *strong manipulator* if he/she is a unique manipulator in the population.

The following theorem asserts that any Pareto rule has a strong manipulator.

Theorem 2.2. *Let ϕ be a Pareto aggregation rule. Then there exists a strong manipulator.*

It is easy to check that any rule which is homotopic to a dictatorial rule has a strong manipulator, but the converse is not ture: the existence of a strong manipulator does not imply that a rule is homotopic to a dictatorial rule.

3. Linear preferences

From now on we study the case of voters with linear preferences. Denote P' be the space of linear preferences on X: this set can be identified with the $(n-1)$-dimensional sphere in $\mathfrak{R}^n$, i.e.,

$$P' = S^{n-1} = \{v \in \mathfrak{R}^n : |v| = 1\}.$$

The space of profiles $(P')^l$ is thus $(S^{n-1})^l$ with l voters.

The main result is the following theorem. It asserts that in the two dimensional case, any Pareto rule is homotopic to a dictatorial one.

Theorem 3.1. *Let $\phi: (S^1)^l \to S^1$ be a continuous Pareto aggregation rule. Then ϕ is homotopic to dictatorial rule.*

Proof. Let us recall the following facts from algebraic topology (Fomenko and Fucks 1989).

1. $S^1 = K(\mathbb{Z}, 1)$. Really, $\pi_1(S^1) = \mathbb{Z}$ and $\pi_i(S^1) = 0\ \forall i \geq 2$.
2. Let X be a CW-complex; let $f: X \to K(\Pi, n)$ be a continuous map ($K(\Pi, n)$ is a space whose n-th homotopy group is Π with all others zero). Assign to a map f the cohomological class $f^*F_\pi \in H^n(X;\Pi)$, where $F_\pi \in H^n(K(\Pi, n); \Pi)$ is the fundamental cohomological class (which corresponds to the identical homomorphism $\mathrm{id}_\pi \in \mathrm{Hom}(H_n(K(\Pi, n), \Pi))$. Here f^* is the map of the cohomology groups which is induced by f (f^*: $H^n(K(\Pi, n);\Pi) \to H^n(X;\Pi)$). Then the map

$$\pi(X, K(\Pi, n)) \to H^n(X; \Pi)$$

$$(f \to f^*F_\pi)$$

is one to one correspondence.

The interested reader can refer to any standard textbook for the proofs of these facts, for example, Fomenko and Fucks, 1989 (Chapter 2.18).

According to 1 and 2, we have the following isomorphism

$$\pi((S^1)^l, S^1) \cong H^1((S^1)^l, \mathbb{Z}).$$

$(S^1)^l$ is the l-dimension torus and $H^1((S^1)^l, \mathbb{Z}) = (\mathbb{Z})^l$ is the free abelian group with l generators. Thus, to each continuous map f there corresponds an l-tuple of integers. Let the tuple $(f_1, \ldots, f_l)$ correspond to the map ϕ. Consider $(S^1)^l$ as the factor-space $\mathfrak{R}^l/\mathbb{Z}^l$ (S^1 is $\mathfrak{R}/\mathbb{Z}$), and define the following map f': $(S^1)^l \to S^1$ by the rule:

$$f'(x_1, \ldots, x_l) = \left(\sum_i f_i \cdot x_i\right) \bmod 1.$$

It is easy to check that the l-tuple $(f_1, \ldots, f_l)$ corresponds to the map f': $(S^1)^l \to S^1$. Therefore, by reference to fact 2 above, f' is homotopic to ϕ.

The unanimity property ensures that the degree of the composition of the diagonal map and ϕ is one, i.e., $\deg(\phi \circ \Delta) = 1$, where

$$\phi \circ \Delta : S^1 \xrightarrow{\Delta} (S^1)^l \xrightarrow{\phi} S^1.$$

The map f' is homotopic to ϕ, therefore $\deg(f' \circ \Delta) = 1$.

Consider the maps $\Delta_i : S^1 \to (S^1)^l$, $i = 1, \ldots, l$, which are defined by the rule: $\Delta_i(x) = (1 - \{x\}, \ldots, 1 - \{x\}, \{x\}, 1 - \{x\}, \ldots, 1 - \{x\}) \in (S^1)^l$, where $x \in \mathfrak{R}$, $\{x\} \equiv x \pmod 1$ ($\{x\} \in \mathfrak{R}/\mathbb{Z}$), $\{x\}$ is in the i-th place and $1 - \{x\}$ in all other places. The Pareto property ensures that $\deg(\phi \circ \Delta_i) \in \{1, 0, -1\}$ holds for all $i = 1, \ldots, l$. Thus, we have

$$\deg(f' \circ \Delta_i) \in \{1, 0, -1\} \quad \forall_i \in \{1, \ldots, l\}.$$

The conditions $\deg(f' \circ \Delta) = 1$ implies $\sum_i f_i = 1$, conditions $\deg(f' \circ \Delta_i) \in \{1, 0, -1\}$ imply $(2f_i - 1) \in \{1, -1\}$, $i = 1, \ldots, l$. That yields the existence of a number k such that $f_k = 1$ and $f_i = 0$ $\forall i \neq k$. Therefore ϕ is homotopic to the dictatorial rule ϕ_k. Q.E.D.

Remark. As an application of this theorem we suggest the following proofs of Theorems 2.1 and 2.2.

Observe that the Pareto property (in the linear case) is equivalent to the following requirement:

$$\phi(p_1, \ldots, p_l) \in \left\{ \sum_{i=1}^{l} \alpha_i p_i \,|\, \alpha_i \geq 0, \quad i = 1, \ldots, l \right\} \cap S^{n-1}. \tag{1}$$

Let $\phi : (S^{n-1})^l \to S^{n-1}$ be a Pareto rule. Consider a plane in $\mathfrak{R}^n$, i.e. a two dimension linear subspace of $\mathfrak{R}^n$. Then the intersection of this plane and S^{n-1} is the circle S^1. According to (1), we can restrict ϕ to this circle. The restriction is a one dimension rule: $\phi' : (S^1)^l \to S^1$.

According to Theorem 3.1, for any $x \in S^1$ there exists a voter k such that

$$\phi(-x, \ldots, -x, \overset{k}{\widehat{x,}} -x, \ldots, -x) = x.$$

Due to the compactness of S^{n-1}, such a number k would be the same for all $x \in S^{n-1}$.

Therefore, in view of *wpac*, for any $t \in [0, 1]$ and $(p_1, \ldots, p_l) \in (S^{n-1})^l$, $t \cdot p_k + (1 - t) \cdot \phi(p_1, \ldots, p_l) \neq 0$. Thus we can define the homotopy by the rule

$$\mathscr{H}(p_1, \ldots, p_l, t) = \frac{tp_k + (1 - t)\phi(p_1, \ldots, p_l)}{|tp_k + (1 - t)\phi(p_1, \ldots, p_l)|}.$$

$\mathscr{H}$ is the desired homotopy between ϕ and the dictatorial rule. This completes the Proof of Theorem 2.1.

The same arguments as before imply that for any $p \in S^{n-1}$ there exists a participant, say the first, such that

$$\phi(p, -p, \ldots, -p) = p.$$

We now check that the first voter is a strong manipulator. Consider a profile $\mathbf{p}_{\{-1\}} \in (S^{n-1})^{\mathrm{L}\setminus\{1\}}$. Calculate the degree of the map $\phi(\cdot, \mathbf{p}_{\{-1\}}) \colon S^{n-1} \to S^{n-1}$. Choose a point $q \in S^{n-1}$ such that $-q \neq p_i \in \mathbf{p}_{\{-1\}}$, $\forall i \in \mathrm{L}\setminus\{1\}$. The map $\phi(\cdot, \mathbf{p}_{\{-1\}})$ is homotopic to the map $\phi(\cdot, q, \ldots, q)$. In fact, the desired homotopy is defined by the rule:

$$\mathscr{H}(t) = \phi\left(\cdot, q, \ldots, q, \frac{\left(t - \frac{i-2}{i-1}\right)q + \left(\frac{1}{l-1} - \left(t - \frac{i-2}{l-2}\right)\right)p_i}{\left|\left(t - \frac{i-2}{i-1}\right)q + \left(\frac{1}{l-1} - \left(t - \frac{i-2}{l-2}\right)\right)p_i\right|}, p_{i+1}, \ldots p_l\right),$$

$$t \in \left[\frac{i-2}{l-1}, \frac{i-1}{l-1}\right), \; i = 2, \ldots, l.$$

The map $\phi(\cdot, q, \dots q)$ is a degree one map, because $\phi(p, q, \dots q)$ maps $(S^{n-1})^2 \to S^{n-1}$ while p and q vary, ϕ is a Pareto rule and $\phi(p, -p, \dots -p) = p$. The same arguments yield that maps $\phi(\mathbf{p}_{\{i\}}, \cdot)$ are homotopic to the rules $\phi(q, \dots q, \overbrace{\cdot}^{i}, q, \dots, q)$, which are degree zero maps. This completes the Proof of Theorem 2.2. Q.E.D.

Corollary 3.1. *Let a map $\phi\colon (S^{n-1})^L \to S^{n-1}$ be a Pareto aggregation rule. Then*

a) *If $\phi(p, \dots, p, \overbrace{-p}^{i}, p, \dots, p) = p$, then $\deg \phi(\cdot, \mathbf{p}_{\{-i\}}) = 0$;*

b) *if $\phi(p, \dots, p, \overbrace{-p}^{i}, p, \dots, p) = -p$, then $\deg \phi(\cdot, \mathbf{p}_{\{-i\}}) = 1$.*

4. Infinite populations

Consider now the case of an infinite (countable) number of voters. Now $P^{\mathbb{N}}$ is the product space endowed with the Tychonov (product) topology.

A map $\phi\colon P^{\mathbb{N}} \to P$ is said to be a Pareto aggregation rule if it is continuous with respect to the product topology and it is a Pareto rule. The following theorem show that Pareto aggregation rules are homologous to dictatorial rules, i.e., a strong manipulator exists for any such rule.

Theorem 4.1. *Let a map $\phi\colon (S^{n-1})^{\mathbb{N}} \to S^{n-1}$ be a Pareto aggregation rule. Then there exists a strong manipulator.*

This Theorem shows the difference between the Pareto rules and rules which are unanimous, anonymous and continuous with respect to Tychonov topology. An anonymous map which is continuous in the Tychonov topology is a constant map (Efimov and Koshevoy 1994), so that there are no maps which are unanimous, anonymous and continuous with respect to Tychonov.

Proof. Pick a vector $p \in S^{n-1}$. Consider the identical profile $\Delta(p) = (p, p, \dots)$ and the following sequence of profiles $\{\Delta_k(p) = (p, p, \dots, \overbrace{-p}^{k}, p, \dots)\}$. According to the Pareto property, we have $\phi(\Delta(p)) = p$, $\phi(\Delta_k(p)) \in \{p, -p\}$. The sequence of profiles $\{\Delta_k(p))\}$ converges to $\Delta(p)$ in the Tychonov topology. Therefore, there exists a number m such that $\phi(\Delta_m(p)) = -p$ and $\phi(\Delta_k(p)) = p$ for all $k > m$. (It is easy to check that the case $\phi(\Delta_k(p)) = p\ \forall k$ is impossible.) Consider the following rule $\phi_{m+1}\colon (S^{n-1})^{m+1} \to S^{n-1}$:

$$\phi_{m+1}(p_1, \dots, p_m, p_{m+1}) = \phi(p_1, \dots, p_{m-1}, p_m, p_{m+1}, p_{m+1} \dots).$$

With respect to the chosen number m and due to Corollary 3.1, we have $\deg \phi_{m+1}(p_1, \dots, p_{m-1}, \cdot, p_{m+1}) = 1$. According to Theorem 2.2, the m-th participant is the strong manipulator for the rule ϕ_{m+1}. Therefore, $\phi(\Delta_k(p)) = p$, $\forall k \neq m$. We now verify that the m-th voter is a strong manipulator for ϕ. Let $p = (p_1, \dots, p_n) \in P^N$ be a profile: let $q \in S^{n-1}$ be such that $q \neq -p_i$, for all $i > m$. Define

$$\mathscr{H}(t) = \phi(\mathrm{H}_1(q, p_1, t), \dots, \mathrm{H}_i(q, p_i, t), \dots),$$

where $H_i(q, p_i, t) \equiv p_i$, $i = 1, \ldots m$,

$$H_i(q, p_i, t) = \begin{cases} \dfrac{\left(t - \left(1 - \frac{1}{2^{i-1}}\right)\right)q + \left(\frac{1}{2^i} - \left(t - 1 + \frac{1}{2^{i-1}}\right)\right)p_1}{\left|\left(t - \left(1 - \frac{1}{2^{i-1}}\right)\right)q + \left(\frac{1}{2^i} - \left(t - 1 + \frac{1}{2^{i-1}}\right)\right)p_1\right|}, & \text{if } t \in \left[1 - \frac{1}{2^{i-1}}, 1 - \frac{1}{2^i}\right), \\ p_1, \text{ if } 0 \le t \le 1 - \frac{1}{2^{i-1}}, \\ q, \text{ if } 1 - \frac{1}{2^i} \le t \le 1, \end{cases}$$

$\mathscr{H}$ is the homotopy between $\phi(\mathbf{p})$ and $\phi_{m+1}(p_1, \ldots, p_m, q)$. This completes the proof. Q.E.D.

5. Classes of continuous rules

Next we discuss conditions that are stronger than continuity, but which ensure that a Pareto rule is homotopic to a dictatorial one. First, consider the space of linear aggregation rules, i.e., rules which are defined as follows

$$\phi(\mathbf{p}) = \frac{\sum_i \alpha_i p_i}{|\sum_i \alpha_i p_i|}, \quad \alpha \in \mathfrak{R}^l_+.$$

A linear aggregation rule is a Pareto rule, and is a continuous rule iff it is homotopic to dictatorial rule. In fact, a linear rule is continuous iff there exists a participant i such that $\alpha_i > \sum_{j \neq i} \alpha_j$. Then this rule is a W-Pareto rule. Therefore, by Theorem 2.1, it is homotopic to a dictatorial rule.

Second, consider the space of Lipschitzian rules. Let P be a metric topological space, ρ be its metric. A rule ϕ: $P^L \to P$ is said be a Lipschitzian rule with a constant C if

$$\rho(\phi(\mathbf{p}), \phi(\mathbf{p}')) \le C \cdot \sum_{i \in L} \rho(p_i, p_i').$$

Consider a spheric metric on S^{n-1}, i.e., the distance between two points is the absolute value of the angle (in radians) between the underlying unit vectors.

Let a rule $\phi: (S^{n-1})^L \to S^{n-1}$ be called an L-rule if it is a Lipschitzian rule with respect to the spherical metric.

Proposition 5.1. *Let a Pareto rule* ϕ: $(S^{n-1})^l \to S^{n-1}$ *be a L-rule with the constant* $C \le 2/(l-1)$. *Then* ϕ *is a* W*-Pareto rule.*

Proof. According to Theorem 2.2, a Pareto rule ϕ has a strong manipulator, say the k-th voter. We will show that for any $p_i \in S^{n-1}$, $i \in L$, we have

$$\phi(p_1, \ldots, p_l) \neq -p_k.$$

The proof is by contradiction. Suppose that there exists a profile $\mathbf{p} = (p_1, \dots, p_l)$ such that $\phi(p_1, \dots, p_l) = -p_k$. Separate the sphere in two half-spheres by the hyper-plane $H = \{\langle x, p_k \rangle = 0\}$.
Consider the following profile $\mathbf{p}' = (p'_1, \dots, p'_l)$ where

$$p'_i = \begin{cases} p_k, & \text{if } \langle p_i, p_k \rangle \geq 0, \\ -p_k, & \text{if } \langle p_i, p_k \rangle < 0. \end{cases}$$

According to (1), there exists at least one voter i such that $\langle p_i, p_k \rangle < 0$. Thus, we have

$$\rho(\phi(\mathbf{p}), \phi(\mathbf{p}')) = \rho(-p_k, p_k) = \pi = \frac{2}{l-1} \cdot (l-1) \frac{\pi}{2} >$$

$$\frac{2}{l-1} \cdot \left(\sum_{i:\langle p_i, p_k \rangle < 0} \rho(p_i, -p_k) + \sum_{i:\langle p_i, p_k \rangle \geq 0} \rho(p_i, p_k) \right).$$

A contradiction. Q.E.D.

Therefore, according to Theorem 2.1, any Pareto L-rule with a constant $C \leq 2/(l-1)$ is homotopic to a dictatorial rule.

Finally, we suggest the following class of Pareto rules. Consider a Pareto rule $\phi\colon (S^{n-1})^L \to S^{n-1}$. Then there exists a voter who is strong manipulator with respect to the rule ϕ, say the i-th voter. A strong manipulator is said to be *super strong manipulator* if for any $p \in S^{n-1}$ the map ϕ^i_p: $(S^{n-1})^{L-1} \to S^{n-1}(\phi^i_p(\mathbf{p}_{-i}) := \phi(p, \mathbf{p}_{-i}))$ is not onto, and the correspondence ξ^i: $S^{n-1} \Rightarrow S^{n-1}$, $\xi^i(p) = S^{n-1} \backslash \operatorname{Im} \phi^i_p$, has a continuous selection (φ: $S^{n-1} \to S^{n-1}$ denotes the selection). For example, any W-Pareto rule has a super strong manipulator. Intuitively, the other voters cannot overcome a super strong manipulator.

If a Pareto aggregation rule ϕ has a super strong manipulator (the i-th voter), then it is homotopic to a dictatorial rule. In fact, the homotopy

$$\mathscr{H}(p_1, \dots, p_l, t) = \frac{-t\varphi(p_i) + (1-t)\phi(p_1, \dots, p_l)}{|-t\varphi(p_i) + (1-t)\phi(p_1, \dots, p_l)|}$$

continuously transforms the rule ϕ into the map $-\varphi\colon S^{n-1} \to S^{n-1}$. According to Corollary 3.1, the map $-\varphi$ is a degree one map. Therefore, $-\varphi$ is homotopic to the identity map, which implies that ϕ is homotopic to the dictatorial rule ϕ_i.

References

Chichilnisky G (1976) Manifolds of preferences and equilibria. Report 27, Harvard University, Cambridge, MA

Chichilnisky G (1982) The Topological equivalence of the Pareto condition and the Existence of a Dictator. J Math Econ 9: 223–233

Chichilnisky G (1993) On strategic control. Q J Econ CVIII: 285–290

Efimov B, Koshevoy G (1994) A topological approach to social choice with infinite populations. Math Soc Sci 27: 145–157

Fomenko A, Fucks D (1989) A course in homotopy topology. Nauka, Moscow (in Russian)

Social choice with infinite populations: construction of a rule and impossibility results

Graciela Chichilnisky[1], **Geoffrey Heal**[2]

[1] Department of Economics, Columbia University, 405 Law Memorial Library, New York, NY 10027, USA
[2] Program on Information and Resources, Columbia University, 405 Law Memorial Library, New York, NY 10027, USA

Received: 30 October 1994/Accepted: 22 April 1996

Abstract. We provide a simple construction of social choice rules for economies with infinite populations. The rules are continuous, Pareto and non-dictatorial; they are constructed as limits of individual preferences when the limit exists, and otherwise as adequate generalizations. This contrasts with the impossibility results of Arrow (1951) and Chichilnisky (1980), which are valid on economies with finitely many individuals. Our social choice rules are, however, limits of dictatorial rules. This paper was written in 1979.

1. Introduction

Social choice seeks ways to aggregate individual preferences into social preferences. The problem is to find maps, called *social choice rules*, from spaces of individual preferences to spaces of social preferences, which satisfy certain desired conditions or *axioms*. We study the problem in the context of large economies, and show that with infinite populations one can construct maps which assign a social preference to each set of individual preferences and satisfy an attractive set of axioms. The maps are continuous, Pareto (which means that if everybody prefers one choice to another then so does the social

---

G. Chichilnisky is at the Department of Statistics, Columbia University, and G.M. Heal was at the Department of Economics, University of Sussex and Columbia University in 1980, and is currently at Columbia Business School. This research was partly supported by the UNITAR project on technology, distribution and North-South relations at the Center for the Social Sciences of Columbia University. The authors thank J.C. Candeal, J. Drèze, C. Fourgeaud, G. Bergman, M. Hirsch, E. Indurain, D. Sonderman for helpful comments.

preference) and non-dictatorial. We also characterize the properties of these rules[1].

The difficulty in constructing social choice rules resides, of course, in the conditions required of them. Three widely used conditions, or *axioms*, are: continuity, anonymity and respect of unanimity[2]. These axioms are best suited for euclidean spaces of choices such as those typically used in economic theory[3]. Because of its affinity with the rest of economic theory, this approach to social choice has led to the discovery of close links between social choice theory and consumer theory[4], and between social choice and classic problems of resource allocation such as the existence of competitive equilibrium and the non-emptiness of the core[5]. Another set of axioms considered for continuous social choice rules consists of the Pareto principle and non-dictatorship.

While these three axioms are simple and appealing, in many cases they may be impossible to satisfy. For example, earlier work by Chichilnisky has shown that any continuous Pareto rule is "almost dictatorial", in the sense that there is an agent who, by appropriate misrepresentation of preferences can obtain any outcome he or she might wish. However, for economies containing infinitely many individuals, there are social choice rules which are continuous, Pareto and non-dictatorial: in this paper we construct such rules in an intuitive fashion, and describe their properties.

We construct a social choice rule as follows. Suppose that an infinite sequence of individual preferences has a limiting preference. Then we simply select that limit as the social preference. This defines a continuous social choice rule which obviously respects unanimity over all such sequences. In particular, this map can be shown to satisfy the Pareto condition. Furthermore this map is not dictatorial.

One then attempts to extend this map to *all* sequences of preferences, whether or not they have limiting preferences. We show that this extension can be achieved while maintaining important properties: the Pareto property and the non-dictatorship properties remain[6]. Therefore, the social choice problem is resolved for economies with infinitely many individuals.

---

[1] This paper was first written in 1979. In recent years its results have been widely circulated, cited and applied, and yet have not been superceded. The following papers have been written building on and using the results of this one: L. Lauwers [28, 29], Efimov and Koshevoy [22], Lauwers and van Liederkerke [30], Candeal, Indurain and Uriarte [5], Candeal, Chichilnisky and Indurain [4], Chichilnisky [16], Chichilnisky [17].

[2] These axioms were introduced in Chichilnisky [7, 8]. They are different from Arrow's [1] axioms for social choice, which are Pareto, independence of irrelevant alternatives and non-dictatorship, although recent work point to deep relations between the two approaches, Baryshnikov [2], also in this issue.

[3] Instead, Arrow's axioms for social choice have been mostly applied to discrete set of choices, and using combinatorial techniques. The main difference in the axioms is that Chichilnisky required continuity while Arrow requires independence from irrelevant alternatives.

[4] See Chichilnisky and Heal [18].

[5] Chichilnisky [11, 13–15].

[6] It is straightforward to show that such a map is Pareto and non-dictatorial. We show that if the point of accumulation is chosen as the limit of a sequence along a free ultrafilter, then this selection can also be done continuously.

The extension to all sequences of preferences is achieved by choosing as a social preference, a point of accumulation of individual preferences. This point of accumulation can be selected in a continuous way but, in general, the result of this selection depends on the order in which the preferences are listed. Hence a permutation of preferences among individuals will affect the social preference, so that the social choice rule, although Pareto and non-dictatorial, is no longer anonymous. Anonymity is a stronger condition than non-dictatorship, a point discussed below. Related results addressing the issue of anonymity for infinite populations have been obtained by Lauwers (this volume), Lauwers and van Liederkirke [30], Candeal et al. [5], Candeal et al. (this volume), Chichilnisky [16, 17], and Effmov and Koshevoy [22].

Infinite populations are an important feature of our results. Our construction of a non-dictatorial rule depends crucially on there being an infinite number of agents: with only a finite number, this rule selects the preference of the last agent, who is then a dictator. Having an infinite population means that there is no "last agent", so one can safely select the limit preference as a social preference without introducing a dictatorship[7]. We show, however, that the social choice rules which we construct are limits of dictatorial rules. In addition, we show that there is a family of individuals of measure as small as desired, which imposes its wishes on society.

2. Notation and definitions

Let X be the *choice space*, $X = R^n$[8]. A *preference* p on X is defined by giving for each choice x in X a preferred direction, a vector denoted $p(x)$, which can be understood as the gradient field of a utility $u: R^n \to R$ representing the preference[9]. A preference is a continuous map $x \to p(x)$ from choices x into vectors in R^n. Such a map is called a *vector field* on X.

As is usual in social choice theory it is the ordering and not the utility values that matter; therefore, the lengths of the vector fields do not matter; one normalizes the vector fields to obtain vectors of unit length, i.e. $\| p(x) \| = 1$ for all x. We topologize the space R^n as usual, and endow the space of preferences with a topology in which the distance between two preferences p, q is the supremum of the euclidean norm of their difference: $\| p - q \| = \sup_{z \in R^n} \| p(x) - q(x) \|$.

---

[7] Other results on social choice with infinitely many agents were obtained by Fishburn, Kirman and Sonderman, and are discussed below.

[8] It suffices to consider a space which is *homeomorphic* to R^n. A topological space X is *homeomorphic* to another Y if there exists an one to one onto continuous function $f: X \to Y$ whose inverse is also continuous.

[9] The vector $p(x)$ is intended to be the normal to the indifference surface when the preference is represented by a utility function $u: R^n \to R$, so that in effect $p(x)$ is the gradient vector $Du(x)$; in other words, the preferences considered here are representable by utility functions with continuous derivatives, also called smooth preferences, see Debreu [20] and Chichilnisky [6].

Definition 1. The space of preferences P is a closed, equicontinuous family of locally integrable continuous unit vector fields on X[10].

The space of continuous bounded vector fields on R^n is an infinite dimensional space endowed with the above topology, and its subspace P contains infinite dimensional manifolds[11]. P also contains finite dimensional spaces of preferences, such as the space of all linear preferences, denoted P_L.

The space P *is compact* by Ascoli's theorem[12].

We assume that there are infinitely many individuals indexed by the set of integers N.

Definition 2. A profile of preferences in P is a sequence of preferences, one for each individual: $\{p_i\}_{i=1,2\ldots} = (p_1, p_2, p_3, \ldots)$: $\forall i$, $p_i \in P$. A profile is therefore an element of the product of the space P with itself indexed by the integers, denoted $\prod_{i=1}^{\infty} P_i$.

P_i denotes the space of preferences of the ith individual and thus $P_i = P$ for all i. A profile is denoted $\{p_i\}_{i=1,2\ldots}$ or simply $\{p_i\}$. If there are finitely many individuals, say, $i=1, \ldots, k$, then the space of profiles is, instead, $\prod_i^k P_i$.

Definition 3. A social choice rule ϕ is a map from profiles in P to social preferences in P. With infinitely many individuals, therefore

sas
$$\phi : \prod_{i=1}^{\infty} P \to P$$

and with k individuals

$$\phi : \prod_{i}^{k} P \to P.$$

Continuity of the social rule ϕ is defined with respect to the topology of the spaces of profiles $\prod_i^k P$ and $\prod_{i=1}^{\infty} P$; the former is endowed naturally with the product topology it inherits from P. A product topology ρ for the spaces $\prod_{i=1}^{\infty} P$ will be defined below.

---

[10] I.e. $p(x) \in P \Rightarrow p(x)$ is locally the gradient of a C^1 real valued function on X, see also Chichilnisky [6] and Debreu [20]. The family P of maps from X into R^n is called *equicontinuous* at $x \in X$ iff there is a neighborhood of x whose image is small under every member of P; formally, P is equicontinuous at x when for every neighborhood V of R^n there is a neighborhood $U(x)$ of X such that $\forall p \in P$, $p(U) \subset V(p(x))$. P is eqicontinuous when it is equicontinous at every x. For P to be equicontinuous it suffices that the vector fields in P are continuously differentiable, with uniformly bounded derivatives.

[11] See Chichilnisky [6]. The space P of all locally integrable preferences is characterized by the Frobenius integrability conditions, which are given by a set of partial differential equations: they are necessary (but not sufficient) conditions for a vector field to be the gradient of a real valued function. Under monotonicity conditions, the conditions are also global, see, for instance, Debreu [10]. The Frobenius conditions and the normalization assumption $\| p(x) \| = 1$ are both closed conditions, so that the space P endowed with the C^1 topology is a complete space since it is a closed subspace of the space of all C^1 vector fields.

[12] See Kelley [27, p. 234].

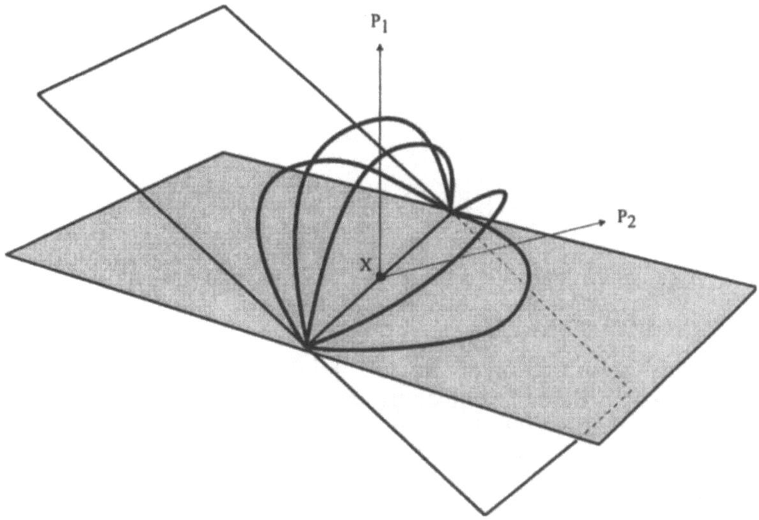

Fig. 1. The Pareto condition for a profile with two different preferences

Definition 4. A social choice rule ϕ is said to satisfy the *Pareto* condition, or simply to be Pareto, if whenever all individuals in the profile $\{p_i\} \in \prod_i P$ prefer choice x to choice y[13] then so does the outcome or social preference assigned to $\{p_i\}$, $\phi(\{p_i\})$.

A simple geometric representation of the Pareto condition is in Lemma 1 below. To motivate it, consider the following example.

Example 5. Fig. 1 represents the case where the choice space is R^3 and there are only 2 individuals with preferences p_1, p_2, defining the profile $\{p_i\}$. If the map $\phi: \prod_i^2 P \to P$ is Pareto then the outcome preference p at a fixed choice x must determine a vector $p(x)$ in the convex cone of vectors having positive inner products with all vectors in the area designated by semi-circles. This is because the Pareto property implies that any utility function representing the outcome preference p at the choice x must increase in the directions that u_i increases, where u_i is a utility representation for the preference p_i, $i = 1, 2$. Therefore $p(x)$ must have positive inner product with all vectors in the shaded area of S^2. These latter vectors are all those that have positive inner product with both p_1 and p_2.

More generally:

Lemma 6. *The map* $\phi: \prod_i P \to P$ *is Pareto if and only if* $\phi(\{p_i\})(x) \in$ *convex hull* $\{p_i(x)\}_{i=1,2\ldots}$[14].

[13] For instance, if the utility representations for the preferences $\{p_i\}_{i=1,2\ldots}$, say $\{u_i\}_{i=1,2\ldots}$, all give higher utility value to x than to y.
[14] The convex hull defined by a set of vectors $\{p_i\}_{i=1,2\ldots}$ is the smallest convex cone of vectors which contains the set $\{p_i\}_{i=1,2\ldots}$. The Pareto condition is vacuously satisfied when the convex hull of $\{p_i]$ is empty.

Proof. See Chichilnisky [10].

Definition 7. A social choice rule $\phi: \prod_i P \to P$ is called dictatorial when there exists an individual $d \in I$ such that $\phi(\{p_i\}) = p_d \; \forall \{p_i\}$ in $\prod_i P$, i.e. when the map ϕ is the projection of the sequence $\{p_i\}$ onto its dth coordinate, ϕ is *non-dictatorial* if there is no d with $\phi(\{p_i\}) = p_d \; \forall \{p_i\} \in \prod_i P$.

Definition 8. A social choice $\phi: \prod_i P \to P$ *respects unanimity* when $\phi(\{p_i\}) = p$ whenever $p_i = p$ for all i.

Definition 9. A social choice rule $\phi: \prod_i P \to P$ is *anonymous* when

$$\phi(\{p_i\}_{i=1,2\ldots}) = \phi(\{p_{k_i}\}_{i=1,2\ldots})$$

where $\{k_i\}$ is any permutation of the set of integers $N = \{1, 2, \ldots\}$. An anonymous rule assigns the same outcome to a set of preferences independently of the order of the voters that have those preferences.

In order to define the social choice map we shall work with a particular compactification of the integers which is used in what follows to select continuously accumulation points in bounded sequences, the Stone–Cêch compactification, denoted βN. The following definitions of filters and their properties can be found in Choquet [19], Kelley [27] and Gillman [24].

Definition 10. A *filter* on a set X is a family F of subsets of X such that
(1) $U \in F$, $V \in F \to U \cap V \in F$
(2) $\phi \notin F$
(3) $U \in F$, $V \supset U \to V \in F$

Definition 11. An *ultrafilter* is a filter F such that there is no other filter $G \neq F$ with $G \supset F$.

A filter on X can be thought of as a family of neighborhoods of points for a topology of X, and an ultrafilter as a maximal such family. By the axiom of choice (Zorn's lemma) any filter is contained in a maximal filter, i.e. in an ultrafilter. An ultrafilter F for a set X is said to be *based on a point* $x \in X$ and is denoted $A(x)$ if it is the equivalent of the set of all neighborhoods of x in X, formally: $A(x)$ is based on the point x if the intersection of all subsets of X belonging to $A(x)$, is the point x.

Definition 12. An ultrafilter F is called *free* if there is no point x in X such tht F is based on x.

Example 13. If N is the set of integers, and B the filter of all subsets in N that are complements of finite sets, then B is called the Frechet filter and is contained in a maximal filter, called the *Frechet ultrafilter*[15]. This ultrafilter is free, i.e. it is not based on any integer in N.

Definition 14. The *Stone–Cêch compactification* of the set of integers N denoted βN is defined pointwise as the set of all ultrafilters of N. N is contained in βN by the identification of each point n in N with the ultrafilter based on n, i.e. the set of all subsets of N that contain the point n, see [27].

---

[15] B is also called the Fréchet filter, Choquet and Marsden [19].

Lemma 15. *The complement of N in βN (i.e. $\beta N - N$) is the set of all free ultrafilters of N. By the definition of the topology of βN, see Kelley [27], N is dense in βN. Therefore, any free ultrafilter of N can be approximated by ultrafilters based on integers in N. Furthermore, any continuous function $f: N \to Z$, where Z is a compact topological space, admits unique extension to a continuous function $\tilde{f}: \beta N \to Z$ which makes the diagram below commutative:*

$$\begin{array}{ccc} \beta N & & \\ i \uparrow & \searrow \tilde{f} & \\ N & \xrightarrow{f} & Z \end{array}$$

In this diagram, $i: N \to \beta N$ is the inclusion of N as a dense subset of βN.

Proof. This is the Stone–Cêch theorem: a proof is in Kelley [27, p. 153]. ■

The following provides another representation for the Sonte Cêch compactification. This definition is useful for our proofs because it explains the properties of "evaluation maps" which will be used in defining social choice rules. The following is a definition of the Stone Cêch compactification used in Kelley [27, p. 152], applied to the special case of the integers, N. Let $C(N)$ denote the space of all functions from the set of integers N into the unit interval I, i.e. a space of sequences of real numbers between 0 and 1. The spaces $I^{C(N)}$ is the product of I taken $C(N)$ times; it consists, by definition, of all functions $\alpha: C(N) \to I$, and it is compact with the product topology[16] by Tychonov's Theorem [27, p. 143].

For any $n \in N$ define the *evaluation map based on* n, $e_n \in I^{C(N)}$, by: $e_n(f) = f(n)$ $\forall f \in C(N)$. Now define the *evaluation map* $e: N \to I^{C(N)}$, as $e(n) = e_n$. Evaluation e is a continuous map of N into the product space $I^{C(N)}$ and defines a homeomorphism of N into a subspace of $I^{C(N)}$.

Definition 16. The closure of the image of N under the evaluation map e, denoted $cl(e[N])$, is a closed subset of the compact space $I^{C(N)}$ and, therefore, a compact space itself: this set $cl(e(N))$ together with the map e define a compactification of the integers N, βN. Since N can be identified with its image under the evaluation map, $e[N]$, it follows by construction that N is dense in βN.

The density of N in βN is used to extend functions from N into βN; this is Stone's theorem, stated above.

A profile of preferences in $\prod_i P$ defines a function from N into P. It follows from Stone's theorem and the fact that P is compact that:

Corollary 17. *A profile of preferences $\prod_i P$ defines a sequence of preferences in P, and therefore it admits a unique continuous extension to a function from ultrafilters to P, i.e. a function on βN to P, called a generalized profile of preferences.*

This result is used in the construction of the social choice rule in the next section.

---

[16] The product topology is the topology in which if $\{\alpha^j\} \subset P^{C(N)}$, $\alpha^j \to \alpha \in P^{C(N)} \Leftrightarrow \forall f \in C(N)$, $\alpha^j(f) \to \alpha(f)$.

Corollary 18. *Consider the space of generalized sequences of preferences $P^{\beta N}$, consisting of all the functions from βN into P, i.e. the product space of P with itself taken βN times. This space is compact.*

Proof. This is Tychonoff's theorem, see e.g. [27]. ■

Definition 19. Endow the space $P^{\beta N}$ with the product topology, namely with the topology of coordinate convergence-: a net $\{f_n, n \in D\}$ converges to $g \Leftrightarrow \{f_n(x), n \in D\}$ converges to $g(x)$ for each x in X[17]. $P^{\beta N}$ contains the infinite sequences of preferences in P, namely P^N, since every integer N is identified uniquely with an ultrafilter in βN.

Definition 20. Let ρ denote the product topology which the space of infinite sequences $P^N \subset P^{\beta N}$ inherits from its inclusion in the space $P^{\beta N}$ with its product topology.

From now on we shall work with the topology ρ on P^N. Note that the topology β is not the standard product topology on P^N; in fact it is finer, since it requires coordinatewise convergence not only on integers but also on all other elements of βN[18].

Lemma 21. *For any $F \in \beta N$, the evaluation map e_F defined as the projection on the coordinate $F \in \beta N$, namely, $e_F(f) = f(F)$ for all $f \in P^{\beta N}$ is a continuous map on $P^{\beta N}$ with its product topology. In particular, e_F defines a continuous map on P^N with the product topology ρ it inherits from its inclusion in $P^{\beta N}$.*

Proof. This is immediate; see e.g. Kelley [27, p. 218, lines 1–7]. ■

3. Construction of social choice rules for infinite populations

In this section we construct a social choice rule for infinite populations. This rule is continuous, Pareto and non-dictatorial: the social preference is the limit of the sequence of individual preferences, or some suitable generalization of the limit when the limit does not exist. Our strategy is as follows. First we establish our results for the simple case of profiles contained in a subset of $\prod_i P$ consisting of preference profiles that have limits with respect to the C^1 topology on vector fields defined above. Denote this by P_l:

$$P_l = \{p_k\} \in \prod_i P_i : \exists p \in P : \| p_k - p \| \to 0 \text{ as } k \to \infty\}.$$

The results are established very easily in this case. Theorem 1 then proves the results for the general case. An outline of the proof of this Theorem is given in this Section, with a complete proof being contained in the appendix.

---

[17] See e.g. Kelley [27, (3.4) and p. 217].

[18] For example, the sequence of preference profiles $\{\alpha_n\}_{n=1,2\ldots}$ defined by $\alpha_1 = (q, p_0, p, p, \ldots) \in P^N$, $\alpha_2 = (q, q, p, p, p, \ldots), \ldots,$ $\alpha_n = (\overbrace{q, q, \ldots, q}^{n-times}, p, p, p, \ldots)$ does not converge to the profile $\alpha = (q, q, \ldots, q, \ldots)$ in the product topology ρ on P^N, even though it does converge to α in the standard product topology of P^N.

We now turn to the case of preference profiles that have limits, i.e. those in P_l. P_l inherits the topology of $\prod_i P$, and is a closed subspace of $\prod_i P$. It is useful to note the following points:

- It follows from the topology on P that if $\{p^n\} \to p$ then $p^n(x) \to p(x)$ for any choice $x \in X$.
- Given the topology of $\prod_i P$, for all choices $x \in X$, $\lim_{m\to\infty}\{p_k\}_{k=1,2,\ldots}^m \to \{p_k\}_{k=1,2\ldots}$ implies that $\forall k = 1, 2, \ldots, \lim_{m\to\infty}\{p_k(x)\}^m \to p_k(x)$ in R^n.
- Even though P_l is a restricted domain for preference profiles, the normals of indifference surfaces of all possible preferences in P_l describe a complete sphere $S^{n-1} \subset R^n$ at each choice $x \in X$.

Lemma 22. *Let $\phi: P_l \to P$ be the social choice rule defined by*

$$\phi(p_1, p_2, p_3, \ldots) = \lim_k \{p_k\}.$$

Then ϕ is continuous, Pareto, non-dictatorial and respects unanimity and $\forall p \in P_l$, $\phi(p)$ defines a complete C^1 locally integrable social preference on the choice space X.

Proof. The map ϕ is well defined by the construction of the space P_l. Continuity of ϕ is immediate given the topology of P_l.

To see that ϕ is Pareto, note that if at any choice x in X the set of vectors $\{p_k(x)\}$ defines a spherical convex hull in S^{n-1}, i.e. if there exists some direction in which all preferences in $\{p_k\}$ increase, then since $\lim_k\{p_k(x)\}$ is contained in the closure of $\{p_k(x)\}$, it is certainly contained in the spherical convex hull of $\{p_k(x)\}$. Therefore by Lemma 1 in Section 2, ϕ is Pareto.

The map ϕ is not dictatorial because if only the dth-coordinate of a sequence in P_l is changed, for any $d = 1, 2, \ldots$, the limit of the sequence does not change. ■

In order to extend the definition of ϕ overall of $\prod_i P$, we use the concept of limit derived from a free ultrafilter on the integers N.

Theorem 23. *There exists a continuous aggregation rule for infinite populations*

$$\psi: \prod_i P \to P$$

which is Pareto and non-dictatorial. ψ is constructed as an extension of the aggregation rule of Lemma 3 *above, and is a limit of dictatorial rules*[19].

Proof. The proof is in the Appendix. Its strategy is as follows. Using the limiting rule defined by a free ultrafilter F of the integers N, we assign to each profile (i.e. to each sequence of preferences in $\prod_i P$) another preference $p \in P$, as follows. The Frechet ultrafilter F assigns to the sequence of preferences $\{p_k\}$ an accumulation point of the set $\{p_k\}$ in P; since P is compact, $\{p_k\}$ has a non-empty set of accumulation points. Therefore the map $\psi\{p_k\})$ is well defined, and it corresponds to the limit of some subsequence of $\{p_k\}_{k=1,2\ldots}$.

[19] The map $\psi \in P^{\beta N}$, i.e., Ψ is in the space of continuous functions from the Stone–Cêch compactification of the integers βN (defined in Sect. 2) into the space of preferences P, and ψ is a limit of dictatorial rules when $P^{\beta N}$ is endowed with the product topology ρ.

By Arzela's theorem [19], this limit defines a continuous vector field. Finally, we show in the Appendix that ψ satisfies the continuity, Pareto, and non-dictatorship conditions, that ψ is a limit of dictatorial rules, and that, if the sequence $\{p_k\}$ has a limit, then $\psi\{p_k\} = \lim_K\{p_k\}$, and therefore ψ coincides with the rule ϕ when restricted to the subset P_l of $\prod_i P$, consisting of sequences that have a limit.

4. Asymptotic dictators with compactified populations

The construction of Theorem 1 shows that the social choice rule is a limit of dictatorial rules; indeed, the proof of Theorem 1 shows that the social choice rule has an "asymptotic dictator", which is an element of the compactified set of individuals. Here we show that the rule we constructed has the property that, when the set of integers is given a finite measure, the asymptotic dictator is "in coalition with" sets of voters of arbitrarily small measure. In this sense, sets of voters of arbitrarily small measure are decisive.

Definition 24. The compactified space of preferences profiles is $P^{\beta N}$ when the space of individuals is the Stone–Cêch compactification of the integers, i.e. βN, as defined in Sect. 2.

Theorem 25. *Assume that the infinite set of voters N is compactified. The rule ψ constructed in Theorem 1 is in this case a dictatorial rule with dictator F, where F is the element of βN used to define ψ, and F is a limit of dictators in the (non-compactified) infinite set of voters N, i.e. it is an asymptotic dictator.*

Proof. In the Appendix.

Definition 26. We say that a subset of individuals $M \subset N$ controls the social choice rule ψ when $\psi(\{p_i\}) = \psi(\{q_i\})$ for every profile $\{q_i\}$ which only differs from $\{p_i\}$ in the preferences of individuals outside the set M, i.e., whenever

$$\forall i \in M, \quad p_i = q_i \Rightarrow \psi(\{p_i\}) = \psi(\{q_i\}).$$

Definition 27. We say that *a dictator D is in coalition with a controlling set of individuals M* if Ψ is dictatorial with dictator D, and Ψ is controlled by the subset M.

In the following we shall assume that the set of voters is given a finite measure μ.

Definition 28. A finite measure on the set of voters is a σ-additive function μ defined on the subsets of the set of integers N, such that $\mu(N) < \infty$.

Corollary 29. *If the space of voters is compactified, then the asymptotic dictator of the continuous Pareto rule ψ is in coalition with controlling sets of voters of arbitrarily small measure.*

Proof. By definition

$$\psi\{p_k\} = ev_F\{p_k\},$$

where F is a free ultrafilter in N, such as the Frechet ultrafilter defined in Sect. 2. Since F is free, it is a limit of a sequence of ultrafilters based on integers, i.e., is based on a sequence of integers $\{j_n\}_{n=1,2\ldots} \to F$ in βN, where the sequence $\{j_n\}_{n=1,2\ldots}$ is unbounded. Therefore by definition, the social choice rule Ψ is controlled by subsets of individuals M^T which are complements of arbitrarily large finite sets of integers $T = \{1, 2, \ldots, T\}$, where $M^T = N - T$. Since the measure μ is finite, $\forall \varepsilon > 0\ \exists T$: $\mu(M^T) < \varepsilon$, thus completing the proof. ■

5. Continuity, independence of irrelevant alternatives, and previous results

The results of the last section provided, for infinitely many voters, a continuous social choice rule that satisfies the Pareto axiom and is non-dictatorial. The construction assigns to each profile of preferences a social preference given by a preferred direction at each point and determining a locally integrable vector field. Therefore, a well defined socially preferred direction is obtained at each choice, and in this sense our aggregate social preference is complete.

Previous work by Fishburn [23] and Kirman and Sonderman [26] studied existence of social choice rules that are Pareto, non-dictatorial and satisfy the axiom of independence of irrelevant alternatives for infinite populations. However, even though both [23] and [26] prove existence results, the social preference whose existence they prove need only be a *weak order*. The problem is that *weak orders* may be incomplete, or even completely trivial. For example, the empty set defines a weak order, so that their results may be empty. Neither [23] or [26] rule out those case so that those results may leave the social choice paradox unresolved in some cases. In our case, the social preference is always complete and not trivial. There is, then, a difference in the "strengths" of the results presented here and those earlier results.

In addition, the approaches of [23] and [26], when applied to continuous choice spaces, may lead to a social preference which is discontinuous, even though all individual preferences are continuous. The demonstration of this point helps to clarify the difference between the approaches. These authors consider a social preference defined by using an ultrafilter as a family of decisive sets. In other words: a choice x is preferred to another y when all individuals in the sets of the ultrafilter prefer x to y.

Fig. 2 depicts preferences at a point x in the choice space, assumed to be a subset of the plane. Each preference is identified by the normal to the indifference surface at x: we show a sequence of preferences $p_1, p_2, p_3, \ldots, p_n, \ldots$, which converge to the preference p_0. As preferences are smooth, the indifference surfaces can be approximated locally by straight lines. So if y is near x, x is preferred to y for every p_i $i = 1, 2, 3, \ldots$, but for p_0, x is indifferent to y.

This construction enables us to bring out the difference between the two approaches. According to the approach of this paper, the social preference at x is clearly p_0. Hence x is socially indifferent to y. However, as x is preferred to y by all individual preferences, there must be a unanimous preference for

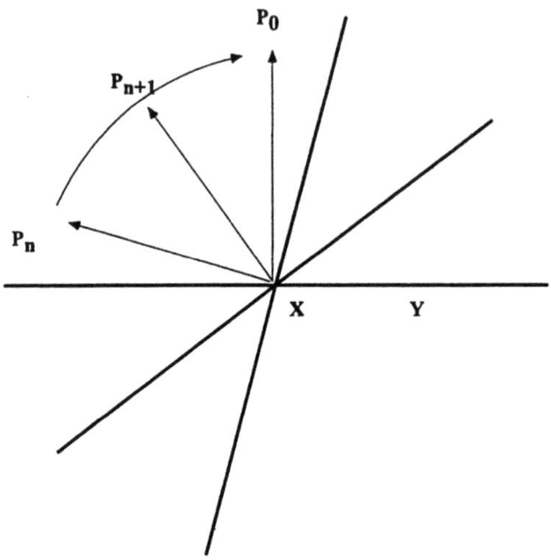

Fig. 2. A sequence P_n of linear preferences converging to P_0. X is preferred to Y by all preferences in the sequence, but for the limiting preference they are indifferent

x over y by some set in any free ultrafilter over the agents, so that according to the approaches of [23] and [26], x is socially preferred to y.

We shall now demonstrate the discontinuity of the social preference derived by using an ultrafilter as a family of decisive sets as done by [23] and [26]. Consider a point z near to and above y. Clearly there exists an $n(z)$, depending on z, such that for all preferences n, $n \geq n(z)$, z is preferred to x. It follows that any free ultrafilter will contain a set of agents who all prefer z to x, so that according to the approach of [23] and [26], z is socially strictly preferred to x. Letting p denote the social preference according to these approaches, we have:

$$z\,p\,x \quad \text{and} \quad x\,p\,y.$$

Now the argument that gives $z\,p\,x$ works for any z above y. Hence we can construct a sequence z_i, converging on y from above, everyone satisfying $z_i\,p\,x$. Thus the z_i satisfy:

$$\lim_{i\to\infty} z_i = y, \quad z_i\,p\,x \text{ all } i, \text{ and } x\,p\,y.$$

It follows that the social preference constructed by [23] and [26] is discontinuous. In particular, it cannot be represented by a gradient vector at x.

Another significant difference of our results with those of [23] and [26] is that our *social choice rule is continuous* on the preferences of the voters, so a small change in individual preferences leads to only a small change in the social preferences. Instead, [23] and [26] require independence of irrelevant alternatives, and do not establish continuity. There is a significant difference between the condition of continuity and the axiom of independence. Independence requires that the outcome *on any pair of choices* be the same if the individual preferences are changed outside of those choices. In order to define the continuity condition, individual preferences must be continuous, and a continuous preference cannot be arbitrarily changed outside of two choices without affecting the preference on (infinitesimally) close choices also.

Therefore, continuity rules out some of the conditions under which independence is tested, since arbitrary changes outside of two choices are not always possible. However, our rule is defined by finding a most preferred direction at each choice, as a function of the most preferred directions and their rates of change, for the individual preferences, at those choices. It follows that outside any small ball of radius say for any $\varepsilon > 0$, around each choice, our rule is independent of irrelevant alternatives. Independence in our framework can, therefore, be defined as follows.

- **Global Independence Axiom:** The outcome of the social rule

$$\psi : \prod_i P \to P$$

at any choice x is the same when the individual preferences in $\prod_i P$ are varied outside of an arbitrarily small neighborhood of x.

And our Theorem 1 in Sect. 3 can now strengthened now as follows:

Corollary 30. *There exists a continuous aggregation rule for infinite populations*

$$\psi : \prod_i \bar{P} \to P$$

which is Pareto, non-dictatorial and satisfies the Global Independence Axiom. ψ *is constructed as in Theorem* 1.

[23] and [26] showed that even though a social choice rule exists, there are still what they term "invisible dictators", so that the resolution of the paradox is more apparent than real in the infinite agent case. Our approach to the problem of social choice with infinitely many agents, has a more straightforward intuitive basis than that of [23] and [26]. Both of these earlier contributions use a non-constructive proof of the existence of a Pareto, non-dictatorial social choice rule: [23] relies on Zorn's Lemma, and [26] use equivalent results. In each case, what is shown is just the existence of a rule: the rule is not characterized.

Instead, here we tackle both the questions of existence and of characterization. Our proof of existence is quite different form, and intuitively more straightforward than, the earlier ones, and also provides a characterization of the kind of rule that it produces. When the rule discussed by [23] and [26] is reinterpreted within the present framework of preferences over Euclidean choice spaces, it turns out to have very different properties from the one discussed here, as pointed out above.

6. Conclusions

We constructed, for infinite populations, social choice rules which are continuous, Pareto and non-dictatorial. Our rules are furthermore globally independent of irrelevant alternatives. These rules were constructed by taking the limit of the sequence of individual preferences, when this exists, and otherwise using an appropriate generalization. It is clear from their construction that these rules may not be anonymous, and that their being non-dictatorial relies in an essential way on there being an infinite population. In fact we showed that these rules are limits of dictatorial rules. For each finite population, the last person is dictator: the problem is solved for infinite populations because in the infinite case, there is no last person.

7. Appendix

This appendix contains the proofs of Theorems 1 and 2 above.

Proof of Theorem 1. Let $\{p_i\}_{i=1,2\ldots}$ be a profile of preferences, i.e. a sequence of preferences in $\prod_i P$; this profile is a map from integers to preferences, and therefore can be written also as a function $f^{\{p_i\}}: N \to P$. By Stone's theorem, stated in Sect. 2, since P is compact, there exists a unique extension of $f^{\{p_i\}}$ to a continuous map $\tilde{f}^{\{p_i\}}$ defined on βN,

$$\tilde{f}^{\{p_i\}}: \beta N \to P.$$

By construction the map $\tilde{f}^{\{p_i\}}$ assigns continuously to each ultrafilter in βN a preference in P, in such a way that if $n \in N$ is identified with the ultrafilter based on n, then $\tilde{f}^{\{p_i\}}(n) = f^{\{p_i\}}(n)$. Continuity refers here to the topology of βN, as defined in Sect. 2.

We now define a social choice rule. Recall the definition of the evaluation map based on the ultrafilter F given in Sect. 2: it maps any function $g: \beta N \to P$ into its value at F, namely $e_F(g) = g(F) \in P$. In particular if $n \in N$ is identified with the ultrafilter based on n, then the evaluation map e_n simply maps the function $g \to e_n(g) = g(n) \in P$.

We use an evaluation map based on a free ultrafilter F of βN, such as for example the Frechet ultrafilter defined in Sect. 2, in order to define the social choice rule. Define:

$$\psi: \prod_{i=1}^{\infty} P \to P \quad \text{as } \Psi(\{p_i\}) = e_F(\tilde{f}^{\{p_i\}}), \tag{1}$$

where the map $\tilde{f}^{\{p_i\}}: \beta N \to P$ in (1) is the unique Stone extension of the map $f^{\{p_i\}}: N \to P$ which defines the preference profile $\{p_i\} \in \prod P$, as defined above.

We now check that the map ψ has all the desired properties. First, continuity. The evaluation map e_F is continuous map on the space of maps $g: \beta N \to P$ into P[20]. Therefore e_F is also continuous on the restriction maps $g/N: N \to P$, which are the profiles of preferences, with the product topology[21].

[20] See Kelley [27, p. 116, Embedding Lemma 4.5]. The continuity is with the product topology.

[21] Kelley [27, p. 152, paragraph 2], considers a topological space X and calls $F(X)$ the family of all continuous functions on X to the closed unit interval Q. The evaluation map e carries a member $x \in X$ into the member $e(x)$ of the cube $Q^{F(x)}$, whose f-coordinate is $f(x)$ for each $f \in F(X)$. He states that "evaluation is a continuous map of X into the cube $Q^{F(X)}$". The topology he uses on $Q^{F(X)}$ is the product topology. In our case $Q = P$, the compact metric space of preferences P, and $X = \beta N$, the Stone Cech compactification of the integers, as defined in Section 2.

Secondly, we prove that the map ψ is Pareto. Since N is dense in βN, the ultrafilter F is a limit in the space βN of ultrafilter based on integers, i.e. $F = \lim_{j\to\infty}\{n^j\}$. By continuity of the evaluation map, $e_F = \lim e_{n^j}$, so that $e_F(f^{\{p_i\}}) = \lim_j e_{n^j}(f^{\{p_i\}}) = \lim_{j\to\infty}\{p_{n^j}\}$, so that $\psi(\{p_i\})$ is an accumulation point of the sequence of vectors $\{p_i\}$. Since $\lim_{j\to\infty}\{p_{n^j}\}$ is an accumulation point of $\{p_i\}$, then, in particular, if the convex hull $\{p_i\}$ is not empty, $\psi(\{p_i\})$ is contained in this convex hull. Therefore by Lemma 1 in Sect. 2, ψ is Pareto.

Next we verify that ψ is not dictatorial; this follows directly from the fact that $e_F \neq e_n$ for any $n \in N$, since the ultrafilter F is a free ultrafilter. We have therefore proved that ψ is continuous, Pareto and non-dictatorial.

Finally, we show that the map ψ is a limit of dictatorial maps. Recall that N is dense in βN, so that the ultrafilter F can be approximated by a net of integers in N: i.e. in the topology of βN, $F = \lim_{j\to\infty}\{n^j\}$. The evaluation map e which maps any function $g: \beta N \to S^{n-1}$ into its value at F, namely $e_F(g) = g(F) \in S^{n-1}$ is continuous, see Kelley [27][22], so that:

$$e_F = e_{\lim\{nj\}} = \lim e_{n^j}.$$

Since e_{n^j} is, by definition, the map that assigns to a sequence of preferences its n^jth coordinate, it follows that e_{n^j} is dictatorial. Therefore, ψ is the limit of dictatorial rules, as we wanted to prove. Finally, since the integers $\{n^j\}$ that approximate F in βN cannot converge to a finite limit (for, otherwise, F would not be free) it follows that when the sequence $\{p_k\}$ has a limit, then $e_{n^j}(\{p_k\}_{i=1,2\ldots})$ converges to $\lim_{k\to\infty}\{p_k\}$. This implies that ψ restricted to P_l coincides with the social choice rule ϕ defined in Theorem 1. ■

Proof of Theorem 2. This is an immediate corollary from Theorem 1, defining the ultrafilter F as an "asymptotic dictator". ■

References

1. Arrow KJ (1963) Social Choice and Individual Values, 2nd ed. Wiley, New York
2. Baryshnikov (1993) Unifying impossibility theorems: a topological approach. Adv App Math 14: 404–415
3. Black D (1948) The rationale of group decision-making. J Polit Econ 56: 23–24
4. Candeal JC, Chichilnisky G, Indurain E (1997) Topological aggregation of preferences: the case of a continuum of agents. Soc Choice Welfare 14: 333–343
5. Candeal JC, Indurain E, Uriarte JR (1992) Some issues related to the topological aggregation of preferences. Soc Choice Welfare 9: 213–227
6. Chichilnisky G (1976) Manifolds of Preferences and Equilibria. Report No. 27, Project on Efficiency Decision Making in Economic Systems. Harvard University
7. Chichilnisky G (1980) Social choice and the topology of spaces of preferences. Adv Math 37(2): 165–176

[22] See Kelley, [27, p. 116, Embedding Lemma 4.5 and p. 152, paragraph 2]: If X is a topological space and $F(X)$ is the family of all continuous functions from X into the closed unit interval Q, the evaluation map e which asigns to each member $x \in X$ the member $e(x) \in Q^{F(X)}$ whose f-coordinate is $f(x)$ for each $f \in F(X)$ is a continuous map from X into $Q^{F(X)}$. In our case, X is the Stone Cech compactification of the integers, βN, defined in Sect. 2, and Q is the space of preferences P.

8. Chichilnisky G (1982) Social aggregation rules and continuity. Q J Econ 1982: 337–352
9. Chichilnisky G (1982) Structural instability of decisive majority rules. J Math Econ 9: 207–221
10. Chichilnisky G (1982) The topological equivalence of the Pareto condition and the existence of a dictator. J Math Econ 9: 223–233
11. Chichilnisky G (1994) Social diversity, arbitrage and gains from trade: a unified perspective on resource allocation. Am Econ Rev 84(2): 427–434
12. Chichilnisky G (1993) On strategic control. Q J Econ 1993: 286–290
13. Chichilnisky G (1995) Limited arbitrage is necessary and sufficient for the existence of a competitive equilibrium. Econ Theory 5(1): 79–108
14. Chichilnisky G (1993) Markets, arbitrage and social choice. [Department of Economics, Columbia University, 1991] CORE Discussion Paper No. 9342
15. Chichilnisky G (1995) A Unified perspective on resource allocation: Limited arbitrage is necessary and sufficient for the existence of a competitive equilibrium, the core and social choice. In: Arrow K, Sen A, Suzamura K (eds) The Theory of Social choice. International Economic Association, McMillan
16. Chichilnisky G (1981) Markets, Games and their Topology
17. Chichilnisky G (1994) What is sustainable development? Stanford Institute for Theoretical Economics Technical Report No. 65
18. Chichilnisky G, Heal GM (1983) Community preferences and social choice. J Math Econ 12: 33–61
19. Choquet G, Marsden J (1969) Lectures on Analysis. Benjamin, Amsterdam, New York
20. Debreu G (1972) Smooth preferences. Econometrica 40: 603–615
21. Dunford N, Schwartz J (1966) Linear Operators (Part I). Interscience, New York
22. Effimov BA, Koshevoy GA (1993) The topological approach to social choice with infinite populations. Math Soc Sciences 27(2): 145–157
23. Fishburn PC (1970) Arrow's impossibility theorem, concise proof and infinite voters. J Econ Theory 2: 103–106
24. Gillman L, Jerison M (1960) Rings of Continuous Functions. Van Nostrand, Princeton, NJ
25. Hirsch MW (1977) Differential Topology. Springer, Berlin Heidelberg New York
26. Kirman A, Sondermann D (1972) Arrow's theorem, many agents and invisible dictators. J Econ Theory 5: 267–277
27. Kelley J (1955) General Topology. (University Series in Higher Mathematics) Van Nostrand, Princeton, NJ
28. Lauwers L (1997) Infinite Chichilnisky rules. Soc Choice Welfare 14: 345–356
29. Lauwers L (1997) A Note on weak ∞-rules. Soc Choice Welfare 14: 357–358
30. Lauwers L, van Liedekerke L (1993) Monotonic Chichilnisky rules with infinite populations. Working Paper, Monitoraat E.T.E.W. Katholiek Universiteit Leuven, Leuven, Belgium
31. Neveu J (1965) Mathematical Foundations of the Calculus of Probability. Holden Day, San Franscisco
32. Spanier E (1966) Algebraic Topology. (McGraw-Hill Series in Higher Mathematics) New York

Topological aggregation, the case of an infinite population

Luc Lauwers

Monitoraat ETEW, KU Leuven, Dekenstraat 2, B-3000 Leuven, Belgium

Received: 30 October 1993/Accepted: 22 April 1996

Abstract. The literature on infinite Chichilnisky rules considers two forms of anonymity: a weak and a strong. This note introduces a third form: bounded anonymity. It allows us to prove an infinite analogue of the "Chichilnisky–Heal-resolution" close to the original theorem: a compact parafinite CW-complex X admits a bounded anonymous infinite rule if and only if X is contractible.

Furthermore, bounded anonymity is shown to be compatible with the finite and the [0, 1]-continuum version of anonymity and allows the construction of convex means in infinite populations. With $X = [0, 1]$, the set of linear bounded anonymous rules coincides with the set of medial limits.

1. Introduction

In order to escape from Arrow's impossibility theorem. Chichilnisky developed the topological approach to the aggregation of preferences.[1] In this framework, a smooth preference on a set $A \subseteq \mathbb{R}^m$ of alternatives is represented by its normalised gradients. This normalisation guarantees that only ordinal information is taken into account. The collection X of these vector fields is a subset of the Banach space of smooth vector fields, and inherits the operator + (the pointwise vector sum) and the C^1 topology. A social

* A former version titled "Monotonic Chichilnisky rules with infinite population" benefited from presentations at the Ethics, Economics and Business Seminar at the Columbia University in New York (March 1993) and the ESEM-conference in Uppsala (August 1993). I would like to thank Luc Van Liedekerke, Bart Capéau, Graciela Chichilnisky, Geoffrey Heal, Esteban Induráin, Jerry Kelly, Gleb Koshevoy and Donald Saari for their comments.

[1] We refer to Chichilnisky (1980, 1982, 1983). In Chichilnisky and Heal (1983) the "resolution of the social choice paradox" is elaborated.

aggregation rule for n individuals, in short an n-rule, is a map

$$\Phi : X^n \to X : (p_1, \dots, p_n) \mapsto \Phi(p_1, \dots, p_n).$$

satisfying continuity with respect to the product topology, anonymity, and a weak form of unanimity. Since the operator + acts upon X, the normalised convex mean Ψ_n defined by

$$\Psi_n(p_1, \dots, p_n) = \frac{p_1 + \cdots + p_n}{\| p_1 + \cdots + p_n \|}$$

comes forward as a candidate to aggregate preferences. However, the condition

$$\Psi_n(p_1, \dots, p_n) \in X \quad \text{for all } p_1, \dots, p_n \in X$$

imposes constraints upon the set X. Indeed, if X coincides with the unit circle, application of Ψ_2 upon the pair $(p, -p) \in X^2$ causes difficulties. But if X is convex, the map Ψ_n is well defined and serves as an (additive) n-rule. In fact, appropriate sets of ordinal preferences structured by means of the sum operator and a topology do admit aggregation rules. The "resolution of the social choice paradox" provides a full characterization of these sets: A parafinite CW-complex X admits an n-rule for all $n \in \mathbb{N}_0$ if and only if X is contractible. In addition, the aggregation rules are shown to be retractions of the convex mean.

The present paper studies infinite aggregation rules, i.e. maps

$$\Phi : X^\infty \to X : \mathbf{p} = (p_1, p_2, \dots, p_n, \dots) \mapsto \Phi(\mathbf{p}).$$

that combine infinite forms of unanimity, continuity, and anonymity. Previous contributions showed that continuity with respect to the uniform topology should be considered.[2] This note concentrates on the role of different versions of the anonymity axiom.

Anonymity aims at denying the impact of the individuals behind preferences and is looked upon as a strong version of non-dictatorship. A symmetry condition, requiring invariance under permutations of the profiles, is frequently used to make this axiom operative. As such, an n-rule does not discriminate between a profile and all of its permutations. Two infinite versions of anonymity appeared in the literature: A weak form, imposing invariance with respect to the group of finite permutations; and, a strong one, demanding invariance with respect to the group of all permutations.[3] We argue that both versions have unappealing features.

First, as finite permutations are unable to transform the profile ${}_{\text{rep}}(p_1, \dots, p_2) = (p_1, p_2, p_1, p_2, \dots, p_1, p_2, \dots)$ into the profile ${}_{\text{rep}}(p_2, p_1) = (p_2, p_1, p_2, p_1, \dots, p_2, p_1, \dots)$, a weak anonymous rule may discriminate between both profiles and therefore treat the sets of odd- and even-indexed individuals unequally. In addition, weak anonymity is compatible with an infinite version of dictatorship: we construct weak anonymous aggregation procedures taking

[2] We refer to Efimov and Koshevoy (1994), Lauwers (1993) and Candeal et al. (1993a).
[3] For example, Chichilnisky and Heal (1979), Candeal et al. (1992). Efimov and Koshevoy (1994) give a general definition of anonymity, but mainly study weakly and strongly anonymous rules.

only arbitrary small fractions of the population into account and completely neglecting the complements of these fractions.

The strong anonymity condition meets the above critiques. Strong anonymity rules out the discrimination between infinite repetitions ${}_{\text{rep}}\mathbf{p}$ and ${}_{\text{rep}}\mathbf{p}'$, whenever $\mathbf{p}, \mathbf{p} \in X^n$ are identical up to a permutation. Also, strong anonymity and dictatorship are incompatible. But again, unacceptable features turn up: strongly anonymous rules are unable to discriminate between, for example, the infinite profiles ${}_{\text{rep}}\mathbf{q}$ and ${}_{\text{rep}}\mathbf{q}'$ with $\mathbf{q} = (p_1, \dots, p_1, p_2)$ and $\mathbf{q}' = (p_1, p_2, \dots, p_2) \in X^n$. Set n equal to 100 and note that profiles in which 1% or 99% of the population hold the preference p_2 are aggregated in the same way. If the difference between p_1 and p_2 is experienced as substantial, this situation is unacceptable.

In view of these difficulties, it is natural to look for rules treating infinite repetitions as finite profiles, i.e. rules Φ for which $\Phi({}_{\text{rep}}\mathbf{p}) = \Psi_n(\mathbf{p})$, where $\mathbf{p} \in X^n$. This requires an anonymity axiom based upon a group of permutations, that allows for discrimination between the infinite repetitions ${}_{\text{rep}}\mathbf{p}$ and ${}_{\text{rep}}\mathbf{p}'$ in exactly those cases where $\Psi_n(\mathbf{p}) \neq \Psi_n(\mathbf{p}')$. The set of bounded permutations, i.e. permutations π for which the ratio $\pi(n)/n$ tends to 1 as n goes to infinity (e.g., the rearrangement $2k - 1 \mapsto 2k$ with $k \in \mathbb{N}_0$ which interchanges odd and even numbers embodies a bounded permutation), appears to be a group of permutations meeting this condition. The corresponding anonymity demand is labeled bounded anonymity. With respect to this version of anonymity the resolution of the social choice paradox is very similar to the original Chichilnisky–Heal theorem: A compact parafinite CW-complex X admits a bounded anonymous aggregation rule if and only if X is contractible.

The next argument indicates that, from a measure theoretical point of view, bounded anonymity takes a place in between the finite and the [0, 1]-continuum version of anonymity. Let us slightly strengthen the interpretation of anonymity, and argue that this axiom aims at denying the impact of the identity of the coalitions behind preferences. It does not matter whether a coalition C or C' holds a particular preference, provided the coalitions C and C' are equally large, where "equally large" is defined by measure. An aggregation procedure will be anonymous if the outcome is invariant under measure preserving permutations of the profile.

In the case of n agents, coalitions are equally large if they have the same cardinality. Since all permutations leave the measure $|C|$ of a coalition C invariant, an anonymous aggregation procedure should be invariant with respect to the set of all permutations.

If the set of agents coincides with the continuum [0, 1], the Lebesgue measure λ is used to display the power of a coalition. A suitable notion of anonymity requires invariance under permutations π on [0, 1] for which $\lambda(\pi(C)) = \lambda(C)$ as C runs over the set of Lebesgue measurable coalitions.[4]

Returning to a countable infinite population, difficulties arise since there is no non-atomic measure on the set of natural numbers. However, it seems natural to assign the same magnitude to, for example, the coalitions $k \cdot \mathbb{N} + 1 = \{kn + 1 \mid n \in \mathbb{N}_0\}$ and $k \cdot \mathbb{N} + 2$. Looking for additive set functions μ such that $\mu(k \cdot \mathbb{N} + i) = 1/k$, a class of purely finitely additive measures

[4] Candeal et al. (1993c) deal with the case of a [0, 1]-continuum of agents.

comes forward. The observation that bounded permutations leave this kind of measures invariant leads to the axiom of bounded anonymity.

Conclude that there are three reasons to develop bounded anonymity: it links finite and infinite rules, it takes its place in between the finite and the [0, 1]-continuum version of anonymity, and it extends the Chichilnisky–Heal resolution result.

The infinite version of the convex mean turns out to be an integral with respect to such a purely finitely additive measure or, in the [0, 1]-continuum case with respect to the Lebesgue measure. Consequently, finite coalitions have measure zero and have therefore no impact on the convex mean. This sharply contrasts with the finite rules Ψ_n which are sensitive in all the coordinates. But, since infinite populations were introduced to model the negligibility of a single agent this insensitivity of infinite rules is not surprising (Aumann 1964; Weiss 1981).

In the special case where X coincides with the unit interval $I = [0, 1]$, a linear map ϕ is a bounded anonymous rule if and only if ϕ satisfies

$$\liminf_n (x_1 + \cdots + x_n)/n \leq \phi(\mathbf{x}) \leq \limsup_n (x_1 + \cdots + x_n)/n.$$

for all $x = (x_1, \ldots, x_n, \ldots) \in I^\infty$. A linear map satisfying the above inequalities is called a "medial limit" (Meyer 1973, uses the term "limit medial"). These medial limits extend the notion of convex mean to the countable infinite population framework.

Section 2 provides notation. Section 3 lists some examples to which the Chichilnisky approach applies. Section 4 studies weakly anonymous rules and shows the incompatibility with a strong form of monotonicity. Section 5 studies monotonic and strongly anonymous rules and axiomatizes the infimum-rule. Section 6 studies bounded anonymity, proves the resolution theorem, and characterizes linear rules. Section 7 briefly discusses the [0, 1]-continuum population framework.

2. Notation

The set of preferences is a space X endowed with a metric d. Thus, X is a Hausdorff topological space. For a profile $\mathbf{p} = (p_1, p_2, \ldots, p_n, \ldots) \in X^\infty$ and a number $k \in \mathbb{N}_0$, $\mathbf{p}^k$ stands for the truncated part $(p_1, p_2, \ldots, p_k) \in X^k$. For a preference p in X, denote by (p) the constant profile $(p, p, \ldots) \in X^\infty$.

The set X^∞ of infinite profiles is equipped with the uniform or sup topology which is induced by the metric d_0 defined by

$$d_0(\mathbf{p}, \mathbf{q}) = \text{supremum}\{d(p_1, q_1), \ldots, d(p_n, p_n), \ldots\},$$

where $\mathbf{p}$ and $\mathbf{q}$ belong to X^∞ (e.g. Jameson 1974).

For a permutation π on $\mathbb{N}_0$ and a profile $\mathbf{p} = (p_1, p_2, \ldots, p_n, \ldots) \in X^\infty$ we write $\pi(\mathbf{p})$ for the profile $(p_{i_1}, p_{i_2}, \ldots, p_{i_n}, \ldots) \in X^\infty$ where $i_k = \pi(k)$. A permutation π is said to be finite if for some $T \in \mathbb{N}_0$, $t > T$ implies $\pi(t) = t$. A permutation π is said to be bounded if $\lim_n \pi(n)/n = 1$.

Definition 1. A map $\Phi: X^\infty \to X$ is said to be (1a) *weakly*, resp. (1b) *bounded*, (1c) *strongly anonymous* if

$$\Phi(\mathbf{p}) = \Phi(\pi(\mathbf{p})),$$

where the profile $\mathbf{p}$ runs over the space X^∞ and π over the set of (1a) finite, resp. (1b) bounded, (1c) all permutations.

Note that strong (resp. bounded) anonymity implies bounded (resp. weak) anonymity. The next definition provides three infinite versions of a Chichilnisky rule.

Definition 2. A map $\Phi: X^\infty \to X$ is said to be (2a) a w-rule, resp. (2b) a b-rule. (2c) an s-rule on X if
i) ϕ is continuous with respect to the uniform topology;
ii) Φ is unanimous, i.e. $\Phi((p)) = p$ for all $p \in X$; and iii) Φ is (2a) weakly, resp. (2b) bounded, (2c) strongly anonymous.

Of course, an s-rule is also a b-rule, and a b-rule is also a w-rule. If X is equipped with an order $\precsim$, the combination of $p \precsim q$ and not $q \precsim p$ is shortened to $p \prec q$. The order $\precsim$ on X extends to X^∞ as follows: for two infinite profiles $\mathbf{p}$ and $\mathbf{q}$, we write $\mathbf{p} \precsim \mathbf{q}$ if and only if for all $k \in \mathbb{N}_0$, we have $p_k \precsim q_k$. The natural order on the interval [0, 1] is denoted by $\leq$.

Definition 3. A map $\Phi: X^\infty \to X$ is called *monotonic*, if
i) $\mathbf{p} \precsim \mathbf{q}$ implies $\Phi(\mathbf{p}) \precsim \Phi(\mathbf{q})$; and
ii) $p \prec q$ implies $(p) \prec (q)$.

Definition 4. *Let* $\Phi: X^\infty \to X$ be a map and let $n \in \mathbb{N}_0$. The map Φ_n defined by

$$\Phi_n: X^n \to X: \mathbf{u} \mapsto \Phi_n(\mathbf{u}) = \Phi({}_{\text{rep}}\mathbf{u}),$$

is called the *induced n-map*.

These induced n-maps Φ_n are unanimous, continuous, and monotonic, when Φ is. If Φ is bounded anonymous, the induced n-maps are anonymous. Section 4 provides an example of a w-rule for which the induced maps are dictatorial and therefore not anonymous.

3. Examples

Besides the original setting of Chichilnisky, the following examples may clarify the discussion. Examples 3 and 4 indicate that the Chichilnisky approach applies to the construction of a preference order over the set of infinitely long vectors. In each example, the set X of preferences coincides with the unit interval I.

Note that the compactness of I implies that infinite profiles in I^∞ have limit points in I. Since a lot of information is absorbed in these limit points, compactness becomes in an infinite population framework an attractive property.

Example 1 (Zhou 1991). Consider a society providing two public goods. The quantities of these goods are given by x and y, and the feasible set is given by

$\{(x, y) | x, y \geq 0$ and $x + y = 1\}$. Individuals have continuous and convex preferences over the goods. The preference of individual i induces a unique maximum (x_i, y_i) upon the feasible set. Suppress the second coordinate and collect the individual preferences in the profile $\mathbf{x} = (x_1, x_2, \ldots, x_n, \ldots) \in I^\infty$. Given a profile, society has to single out an allocation. Choosing an appropriate set X of single peaked preferences over the feasible set, parameterised by the peaks, this voting problem boils down to the aggregation of preferences. In the finite population case, the convex mean $\Psi_n(x_1, \ldots, x_n) = (x_1 + \cdots + x_n)/n$ comes forward as an n-rule. This map is continuous, anonymous, unanimous, monotonic, and neutral with respect to the two goods, i.e. the outcome does not change when the x_k are replaced by the $y_k = 1 - x_k$.

Example 2 (Chichilnisky 1980). Let $A \subseteq \mathbb{R}^2$ and let Y be the set of linear preferences over A. Since for an element p of Y the normalized gradient field is constant, the set Y coincides with the unit circle in $\mathbb{R}^2$. The subset Z of the monotonic linear preferences coincides with the intersection of this unit circle with $(\mathbb{R}^+)^2$. The parameterization defined by

$$(\cos(t \cdot \pi/2), \sin(t . \pi/2)) \mapsto t$$

is a homeomorphism of Z and the unit interval I. Here, the preference represented by $t = 0$ indicates neutrality with respect to the second good, i.e. $t = 0$ implies that the choice between the alternatives (x, y) and (x', y') does not depend upon the values of y and y'.

Example 3 (Lehrer 1989; Güth et al. 1991). Let $x \in I$ be the one-shot-game payoff of a player in a game G. Then, a sequence $(x_1, x_2, \ldots, x_t, \ldots) \in I^\infty$ where $t \in \mathbb{N}_0$ is a time index, represents the payoff of the infinitely repeated game G^∞. In developing a strategy, a player orders the set of payoff sequences. It is natural to assume that (*i*) $\mathbf{x} \geq \mathbf{y}$ implies that $\mathbf{x}$ is preferred to $\mathbf{y}$; and (ii) $x > y$ implies that (x) is strictly preferred to (y). The combination of this monotonicity condition with continuity results in a continuous representation Φ of the ordering satisfying $\Phi((x)) = x$ for all $x \in I$ (e.g. Lauwers 1993). Note that this property "formally" coincides with unanimity. Finally, imposing anonymity excludes impatient evaluations.

Conclude that, within this interpretation, an infinite rule on I refers to a monotonic, continuous and time-neutral evaluation on the set I^∞ of payoff sequences. Time averaging, i.e. the map defined by $\mathbf{x} \mapsto \lim\inf_n(x_1 + \ldots + x_n)/n$, is an example of such an evaluation.

Example 4 (Diamond 1965; Campbell 1985). Consider a constant population economy where each generation lives for one period and let $x_t \in I$ denote the consumption level of generation t. Infinite horizon planning problems presuppose an ordering upon the set I^∞ of consumption paths. As in the previous example, imposing monotonicity, continuity, and equity towards future generations (i.e. anonymity) results in a representation by an infinite rule upon I. Equity is imposed to derive a social preference relation without recourse to discounting future consumption levels.

Observe that the framework in Examples 3 and 4 differs from the original setting. Indeed, the Chichilnisky aggregation procedure only considers ordinal information of the individual preferences. And, the ranking of payoff sequences (Example 3) or consumption paths (Example 4) when interpreted as an aggregation procedure, involves cardinality. In particular, it is assumed that each "time index" or "generation" $k \in \mathbb{N}_0$ is equipped with a preference order represented by the utility function $u_k(x) = x$. The aggregated ordering depends upon this particular utility representation (see Lauwers 1994 for details on cardinality and intertemporal comparibility of utilities in an infinite framework). In Sect. 5 we return to this item.

4. Weak anonymity

In case the space X of preferences is compact, one can construct a w-rule as follows. Let $\mathscr{F}$ be a free ultrafilter on $\mathbb{N}_0$. Define a map $\lim_{\mathscr{F}}: X^\infty \to X$ by $\lim_{\mathscr{F}}(\mathbf{p}) = q$, where q is the unique cluster point of the profile $\mathbf{p}$ for which there exists a set of indices $(i_1, i_2, \ldots, i_k, \ldots) \in \mathscr{F}$ such that the sequence $p_{i_1}, p_{i_2}, \ldots, p_{i_k}, \ldots$ converges to q. This map is a w-rule. If X is equipped with the operator $+$, then $\lim_{\mathscr{F}}$ is also additive; i.e. for any pair $\mathbf{p}, \mathbf{q}$ of profiles, we have

$$\lim_{\mathscr{F}} (\mathbf{p} + \mathbf{q}) = \lim_{\mathscr{F}} (\mathbf{p}) + \lim_{\mathscr{F}} (\mathbf{q}).$$

That compactness is sufficient for the existence of topological aggregators is experienced as a paradox (Efimov and Koshevoy 1994). Where contractibility is necessary to construct finite rules, this condition disappears in an infinite framework. The weak anonymity axiom is responsible. This axiom allows for aggregators based upon ultrafilters, which are interpreted as dictators (Kirman and Sondermann 1972; Schmitz 1977). Indeed, for any $k \in \mathbb{N}_0$, there is exactly one number $1 \leq i_k \leq k$ for which the coalition $k \cdot \mathbb{N} + i_k$ belongs to $\mathscr{F}$. Although this coalition represents only a fraction of $1/k$ of the total population, it "dictates" the aggregate. A glance upon the induced maps gives additional evidence. The previous argument implies that the induced map $(\lim_{\mathscr{F}})_k : X^k \to X$ coincides with the projection upon the i_kth coordinate, i.e. agent i_k is a dictator.

If X is equipped with an order, an additional condition upon the aggregator might reflect the extra structure. The following monotonicity axiom can be interpreted as a sensitivity condition.

Definition 5. Let $(X, \preceq)$ be an ordered set, let $\Phi: X^\infty \to X$ be a map, and let $n \in \mathbb{N}_0$. Then, Φ is said to be n-monotonic if $\Phi(p) \prec \Phi(q)$, whenever $\mathbf{p} \preceq \mathbf{q}$ and the coalition $\{i \mid p_i \prec q_i\}$ has a cardinality greater than n.

In the public good example, this axiom requires that when n agents shift their peaks in favour of the first public good, the aggregated allocation shifts in the same direction. In the construction of a preference order over the set of consumption paths, n-monotonicity means that an increase in the consumption of n time periods results in a strictly higher ranked path. The following proposition shows that the combination of continuity and weak anonymity is incompatible with n-monotonicity. The proof closely follows Diamond (1965, p. 176).

Proposition 1. *Let Φ be a w-rule upon a convex subset X of a normed vector lattice. Then, Φ is not* n*-monotonic.*

Proof. Fix two points $p \preceq q$ in X. Parameterize the convex combination $\alpha.p + (1-\alpha).q$ by $\alpha \in [0, 1]$. Within this notation, consider the profile

$$\mathbf{q}_0 = (n[0, 1]; {}_n[0, \tfrac{1}{2}, 1]; \dots ; {}_n[0, 1.2^{-k}, 2.2^{-k}, 3.2^{-k}, \dots, (2^k - 1).2^{-k}, 1]; \dots)$$
$$\uparrow \qquad\qquad\qquad\qquad\qquad\qquad \uparrow$$

where ${}_n$ indicates that the subsequent vector is repeated n times, and the profile

$$\downarrow \qquad\qquad\qquad\qquad\qquad\qquad \downarrow$$
$${}_k\mathbf{q} = (n[1, 1]; {}_n[0, \tfrac{1}{2}, 1]; \dots ; {}_n\underline{[0, 1.2^{-k}, 2.2^{-k}, 3.2^{-k}, \dots, (2^k - 1).2^{-k}, 0]}; \dots).$$

obtained from $\mathbf{q}_0$ by interchanging the initial O with the indicated 1. From ${}_k\mathbf{q}$ we obtained the profile

$$\mathbf{q}_k = (n[1, 1]; {}_n[0, \tfrac{1}{2}, 1]; \dots ; {}_n\underline{[0, 0, 1.2^{-k}, 2.2^{-k}, 3.2^{-k}, \dots, (2^k - 1).2^{-k}]}; \dots).$$

and this by a finite number of interchanges. Since Φ is w-anonymous, we have $\Phi(\mathbf{q}_0) = \Phi(\mathbf{q}_k)$. The sequence $\mathbf{q}_0, \mathbf{q}_1, \dots, \mathbf{q}_k, \dots$ converges to the profile

$$\mathbf{q} = (n[1, 1]; {}_n[0, \tfrac{1}{2}, 1]; \dots ; {}_n[0, 1.2^{-k}, 2.2^{-k}, 3.2^{-k}, \dots, (2^k - 1).2^{-k}, 1]; \dots).$$

Continuity implies that $\Phi(\mathbf{q}_0) = \lim_k \Phi(\mathbf{q}_k) = \Phi(\lim_k \mathbf{q}_k) = \Phi(\mathbf{q})$. Since, $\mathbf{q}_0 \leq \mathbf{q}$ and the cardinality of $\{i \in \mathbb{N}_0 \mid \mathbf{q}_{0_i} \neq \mathbf{q}_i\}$ is equal to n, the map Φ is not n-monotonic.

5. Strong anonymity

As already indicated in the introduction, the strong anonymity condition is a severe demand. The next proposition shows that imposing s-anonymity prohibits the convex mean from extending to an infinite population.

Proposition 2 (Candeal et al. 1992, Theorem 3). *Let X be a convex set of preferences equipped with the sum operator. Let $\Phi: X^\infty \to X$ be an unanimous and s-anonymous map. Then Φ is not additive.*

This proposition indicates that in order to obtain an s-rule, the sum operator on X, as in the original Chichilnisky framework, is redundant. One has to consider other kinds of structures. Efimov and Koshevoy (1994) propose to study s-rules in a complete topological semilattice. This concept (e.g. Peressini 1967) refers to an ordered topological space $(X, \preceq)$ for which the maps

$$\inf: X^\infty \to X : \mathbf{p} \mapsto \text{infimum}_k\{p_1, p_2, \dots, p_k, \dots\}$$

$$\sup: X^\infty \to X : \mathbf{p} \mapsto \text{supremum}_k\{p_1, p_2, \dots, p_k, \dots\}$$

are continuous. Note that such a structure turns both these maps into s-rules on X.

Efimov and Koshevoy (1994, Theorem 6) proved that an s-rule Φ is the limit of its induced rules, i.e. $\Phi(\mathbf{p}) = \lim_k \Phi_k(\mathbf{p}^k)$ for all $\mathbf{p}$, if and only if the aggregate $\Phi(\mathbf{p})$ only depends upon the closure of the set $\{p_1, \dots, p_n, \dots\}$. The following proposition sharpens this result.

Proposition 3. *Let Φ be a monotonic s-rule on a complete topological semilattice $(X, \preceq)$. Then, $\Phi(\mathbf{p}) = \lim_k \Phi_k(\mathbf{p}^k)$ for all infinite profiles $\mathbf{p}$ if and only if $\Phi(\mathbf{p}) = \Phi_2(\inf(\mathbf{p}), \sup(\mathbf{p}))$, where Φ_2 is the 2-rule induced by Φ.*

Proof. First, the induced maps Φ_n of a monotonic, s-anonymous and unanimuous map Φ are studied. Let $\mathbf{q}$ be a vector in X^n. Denote $\inf\{\mathbf{q}\}$ by q_* and $\operatorname{supp}\{\mathbf{q}\}$ by q^*. Notice that there are permutations π_1 and $\pi_2 \in \pi_s$ such that

$$\pi_1({}_{\text{rep}}(q_*, \dots, q_*, q^*)) = \pi_2({}_{\text{rep}}(q_*, q^*, \dots, q^*)) = {}_{\text{rep}}(q_*, q^*).$$

where the vectors $(q_*, \dots, q_*, q^*)$ and $(q_*, q^*, \dots, q^*)$ are of length n. From s-anonymity and the definition of the induced maps, we have

$$\Phi_n(q_*, \dots, q_*, q^*) = \Phi_n(q_*, q^*, \dots, q^*) = \Phi_2(q_*, q^*).$$

As any vector $\mathbf{q} \in X^n$, after a permutation, satisfies

$$(q_*, \dots, q_*, q^*) \preceq \mathbf{q} \preceq (q_*, q^*, \dots, q^*).$$

the monotonicity axiom implies that $\Phi_n(\mathbf{q}) = \Phi_2(q_*, q^*)$.

Now, for an infinite profile $\mathbf{p}$, the sequence $\inf\{\mathbf{p}^2\}, \dots, \inf\{\mathbf{p}^n\}, \dots$ converges to $\inf\{\mathbf{p}\}$ and $\sup\{\mathbf{p}^2\}, \dots, \sup\{\mathbf{p}^n\}, \dots,$ to $\sup\{\mathbf{p}\}$. Since Φ_2 is continuous and Φ is the limit of the sequence of its induced rules, the "only if" part of the proposition follows. The "if" part is easy to check.

Dealing with the evaluation of consumption paths (Example 4) the approximation condition $\Phi = \lim_k \Phi_k$, reflects the idea that in evaluating infinite sequences one falls back on comparing finite parts. Therefore, this condition may be interpreted as a consistency demand. Let us consider the case $X = I$. s-rules satisfying the conditions of the previous proposition can be further specified.

Corollary 1. *Let Φ be a monotonic s-rule on I, such that Φ is approximated by its induced rules. Let Φ be decomposable, i.e. for a sequence x and an infinite subsequence y of x, we have*

$$\Phi(\mathbf{x}) = \Phi(\mathbf{x}'),$$

where $\mathbf{x}'$ *is obtained from* $\mathbf{x}$ *after replacing the* y_i *by* $\Phi(y)$.[5] *Then,*

$$\text{either } \Phi(\mathbf{x}) = \inf\{\mathbf{x}\} \text{ or } \Phi(\mathbf{x}) = \sup\{\mathbf{x}\}.$$

Proof. The assumptions imply that $\Phi(x) = F\{\inf\{\mathbf{x}\}, \sup\{\mathbf{x}\})$ with $F = \Phi_2$. Consider the profile ${}_{\text{rep}}(x, y, u, v)$. Apply decomposability and obtain

$$F[F(x, y), F(u, v)] = F[F(x, u), F(y, v)].$$

In case $F(\cdots)$ is sensitive in both arguments, it follows that F is of the form

$$F(x, y) = f\left(\frac{f^{-1}(x) + f^{-1}(y)}{2}\right),$$

where f is a continuous and strictly monotonic function (Aczèl 1966, Sect. 4.1). Now, supply the decomposability property upon the profile ${}_{\text{rep}}(x, x, y)$. It follows that $F(x, F(x, y)) = F(x, y)$. This contradicts with the functional form of F. Hence, F is not sensitive in both arguments.

Starting from Corollary 1, the Rawlsian infimum-rule can be isolated by imposing an additional condition: an infinite version of Hammond's equity

[5] This definition of decomposability appears in Efimov and Koshevoy (1994).

principle (e.g. Sen 1986). This axiom expresses that a decrease in a low value of a profile cannot be compensated by an increase in a high value.

Definition 6. A rule Φ on I is said to satisfy Hammond's equity principle, if for two profiles $\mathbf{x}$ and $\mathbf{y}$ identical up to two coordinates, say r and t, for which $x_r > y_r \geq y_t > x_t$, we have $\Phi(\mathbf{y}) \geq \Phi(\mathbf{x})$.

Corollary 2. *Let Φ be a decomposable and monotonic s-rule on I, approximated by its induced rules. Let Φ satisfy Hammond's equity principle. Then,* $\Phi(\mathbf{x}) = \inf\{\mathbf{x}\}$.

Note that, the infimum rule when used to rank consumption paths generates large indifferences classes, since it focusses on the poorest generation.

Now, let us return to the observation that the evaluation of consumption paths requires cardinal information and intergenerational comparable utilities. Consider the stream $\mathbf{x} = (x_1, x_2, \dots, x_n, \dots)$ and a transformation $\vartheta(\mathbf{x}) = (\vartheta_1(x_1), \vartheta_2(x_2), \dots, \vartheta_n(x_n), \dots)$ where the ϑ_k are positive monotonic transformations. Both utility functions $u_k(x_k) = x_k$ and $\vartheta_k(x_k)$ represent the same (ordinal) preference relation on the set I. Nevertheless, the Rawlsian infimum rule may change the ranking of a pair x, y after a transformation into $\vartheta(\mathbf{x})$, $\vartheta(\mathbf{y})$.

6. Bounded anonymity

Let Φ be a bounded anonymous rule upon a set X of preferences. Then, for any number $n \in \mathbb{N}_0$, the induced map Φ_n is an n-rule on X. Application of the Chichilnisky–Heal theorem results in the "only if" part of the following infinite version of the resolution theorem.

Theorem 1. *A compact parafinite CW-complex X admits a bounded anonymous infinite rule if and only if X is contractible.*

Proof. Start with a convex subset X of a Banach space. The convex mean, given by

$$\Psi_n(p_1, \dots, p_n) = \frac{1}{n} \cdot (p_1 + \dots + p_n)$$

is well defined upon X^n, for every $n \in \mathbb{N}_0$. Let $\mathbf{p} = (p_1, p_2, \dots, p_n, \dots)$ be an infinite profile, and write $|\mathbf{p}|^n$ for $\Psi_n(p_1, \dots, p_n)$. Consider the infinite sequence

$$\Psi_\infty(\mathbf{p}) = (|\mathbf{p}|^1, |\mathbf{p}|^2, \dots, |\mathbf{p}|^n, \dots).$$

We claim that the map $\Psi_{\mathscr{F}}: X^\infty \to X: \mathbf{p} \mapsto \lim_{\mathscr{F}}[\Psi_\infty(\mathbf{p})]$, where $\lim_{\mathscr{F}}$ is constructed by means of a free ultrafilter $\mathscr{F}$ as in Sect. 4, is a bounded anonymous rule upon X. Since X is assumed to be compact, the map $\Psi_{\mathscr{F}}$ is well defined. Let the sequence $|\mathbf{p}|^{i_1}, |\mathbf{p}|^{i_2}, \dots, |\mathbf{p}|^{i_k}, \dots$ converge to q and let π be a bounded permutation on $\mathbb{N}_0$. It suffices to show that the sequence $|\pi(\mathbf{p})|^{i_1}, |\pi(\mathbf{p})|^{i_2}, \dots, |\pi(\mathbf{p})|^{i_k}, \dots$ converges to the same limit. Let f be defined by $f(n) = [\mathrm{Max}_{i \leq n} |\pi(i) - i|]/n$. Then, the following holds:

$$\| |\mathbf{p}|^{i_k} - |\pi(\mathbf{p})|^{i_k} \| \leq 2.f(i_k) \cdot \| p \|,$$

with p the maximizer of the norms $\| \mathbf{p}_i \|$ where i runs over $\mathbb{N}_0$. Since π is bounded, $f(n)$ tends to zero as n goes to infinity. Hence, the sequence $|\pi(\mathbf{p})|^{i_k}$ also converges to q. Conclude the theorem.

The construction of the rule $\Psi_{\mathscr{F}}$ shows that for any $n \in \mathbb{N}_0$ the induced n-rule $(\Psi_{\mathscr{F}})_n$ coincides with the convex mean Ψ_n. Secondly, the map $\Psi_{\mathscr{F}}$ depends upon the choice of the free ultrafilter $\mathscr{F}$. Hence, the abundance of free ultrafilters on $\mathbb{N}_0$ implies (*i*) the existence of at least as many bounded anonymous aggregation rules, and (*ii*) that the relationship between a bounded anonymous rule and its corresponding sequence of induced n-rules is not a 1–1 relation. Third, the rule $\Psi_{\mathscr{F}} = \lim_{\mathscr{F}} \circ \Psi_\infty$ is the composition of two linear maps, and thus linear.[6] Finally, if X is equipped with an order, the map $\mathbf{p} \mapsto \liminf_n(\Psi_\infty(\mathbf{p}))$ is an example of a nonlinear b-rule.

The rest of this section deals with b-rules on I. First, we notice that bounded anonymity conflicts with n-monotonicity, and this without any continuity requirement (compare with Proposition 1).

Proposition 4. *A map $\Phi: I^\infty \to I$ cannot combine b-anonymity with n-monotonicity.*

Proof. Notice that the profiles $\mathbf{x} = (0, \dots, 0, 1, 0, 1, \dots, 0, 1, \dots)$ and $\mathbf{y} = (1, \dots, 1, 0, 1, \dots, 0, 1, \dots)$ with the first 1, resp, 0, at the $(n+1)$th place are identical up to a bounded permutation. B-anonymity requires $\Phi(\mathbf{x}) = \Phi(\mathbf{y})$, while n-monotonicity requires $\Phi(\mathbf{x}) < \Phi(\mathbf{y})$. A contradiction.

Now let Φ be a linear b-rule upon I. Then, Φ is monotonic. Also, Φ induces a measure μ upon the family $2^{\mathbb{N}_0}$ of subsets of $\mathbb{N}_0$, defined by $\mu(C) = \Phi(\mathbf{c})$ where $\mathbf{c}$ is the characteristic map of C, i.e. $c_k = 1$ or 0 as $k \in C$ or $k \notin C$. The properties of Φ imply that $\mu(k \cdot \mathbb{N}_0) = 1/k$ for all $k \in \mathbb{N}_0$ and that μ is purely finitely additive, i.e. $\mu(C) = 0$ if C is a finite coalition. Denote l_∞ for the set of bounded sequences. Yosida and Hewitt (1952, Theorem 2.3) established a 1–1 relationship between the set of linear maps on l_∞ and the set of measures on $\mathbb{N}_0$. Hence, the study of the linear b-rule Φ reduces to the study of the induced measure μ. The integral Φ with respect to the measure μ is said to be purely finitely additive, since μ is.

In addition, Yosida and Hewitt showed that for every continuous purely finitely additive map $f: l_\infty \to \mathbb{R}$ there exists a finite Borel measure φ on the space Ω of free ultrafilters on $\mathbb{N}_0$ such that

$$f(\mathbf{x}) = \int_\Omega \lim_{\mathscr{F}}(\mathbf{x}) \, \mathrm{d}\varphi(\mathscr{F}),$$

for all $\mathbf{x} \in l_\infty$. Hence, purely finitely additive maps are exactly those linear maps that only take the limit points of a profile into account.

Let C and C' be two coalitions with characteristic maps $\mathbf{c}$ and $\mathbf{c}'$. Then, if C and C' are identical up to a bounded permutation, the sequences $\Psi_\infty(\mathbf{c})$ and $\Psi_\infty(\mathbf{c}')$ have for every free ultrafilter the same accumulation point. Hence, for a b-anonymous measure μ, we have that $\mu(C)$ only takes the limit points of $\Psi_\infty(\mathbf{c})$ into account. Consequently, also the integral Φ with respect to μ is

[6] Candeal and Induráin (1993b) study algebraic aggregation procedures. Their Theorem 3 shows that only the trivial group admits strongly anonymous algebraic rules. The additivity of $\Psi_{\mathscr{F}}$ suggests the existence of bounded anonymous algebraic rules on non-trivial groups.

based upon the limit points of $\Psi_\infty(\mathbf{x})$ with $\mathbf{x}$ a profile. Apply the representation theorem to obtain the following result.

Proposition 5. *Let Φ be a linear b-rule on the unit interval I. Then, there is a finite Borel measure φ on the space Ω of free ultrafilters on $\mathbb{N}_0$ such that*

$$\Phi(\mathbf{x}) = \int_\Omega \lim_{\mathscr{F}} [\Psi_\infty(\mathbf{x})] \, d\varphi(\mathscr{F}).$$

Conversely, any map Φ of the above form defines a linear b-rule.

This proposition indicates that the maps $\Psi_{\mathscr{F}} = \lim_{\mathscr{F}} \circ \Psi_\infty$ and their linear combinations are the only linear *b*-rules upon the unit interval. The next theorem connects the set of linear *b*-rules with the set of medial limits.

Theorem 2. *A map $\Phi: I^\infty \to I$ is a linear b-rule if and only if Φ is a medial limit, i.e. Φ is linear and satisfies*

$$\liminf_n \frac{x_1 + \ldots + x_n}{n} \leq \Phi(\mathbf{x}) \leq \limsup_n \frac{x_1 + \cdots + x_n}{n},$$

for all $\mathbf{x} \in I^\infty$.

Proof. Note that the inequalities

$$\liminf_n [\Psi_\infty(\mathbf{x})] \leq \lim_{\mathscr{F}} [\Psi_\infty(\mathbf{x})] \leq \limsup_n [\Psi_\infty(\mathbf{x})]$$

hold for all $\mathbf{x} \in I^\infty$ and $\mathscr{F} \in \Omega$. Hence, the "only iff" part follows from Proposition 5 and the fact that integration preserves inequalities.

Now, let Φ be a medial limit. Then, Φ is continuous, unanimous and extends by linearity to the set l_∞ of bounded sequences. Let π be a bounded permutation and let $\mathbf{x} \in I^\infty$ be a profile. With $f(n)$ given by $[\mathrm{Max}_{i \leq n} |\pi(i) - i]/n$ we have that $-f(n) \leq |\pi(\mathbf{x}) - \mathbf{x}|^n \leq f(n)$. Since $\lim_n f(n) = 0$, we have $\liminf_n |\pi(\mathbf{x} - \mathbf{x}|^n = \limsup_n |\pi(\mathbf{x}) - \mathbf{x}|^n = 0$. This implies that $\Phi(\pi(\mathbf{x}) - \mathbf{x}) = 0$.

We end this section by observing that for a medial limit or a linear *b*-rule Φ, we have

$$\Phi(x_1, x_2, \ldots, x_t, \ldots) = \Phi(x_2, x_3, \ldots, x_t, \ldots),$$

for all profiles $\mathbf{x}$. Conclude that medial limits are examples of Banach generalised limits.

7. A continuum of agents

Now, we consider the continuum $I = [0, 1]$ of agents, equipped with the Lebesgue measure λ, and study aggregation rules in the special case where X coincides with the unit interval (see Candeal et al. 1993c for a more general framework). Note that, since only measurable sets are considered, some coalitions are excluded. A continuum profile $\mathbf{p}$ is a λ-measurable map $\mathbf{p}: I \to X: t \mapsto \mathbf{p}(t)$. Denote $\mathscr{L}(I, X)$ the set of λ-integrable profiles. A permutation π is said to be λ-invariant if $\lambda(C) = \lambda(\pi(C))$ for all measurable coalitions C.

Definition 7. A continuum rule on X is a map $\Phi:\mathscr{L}(I,X)\to X$ that satisfies unanimity; anonymity, i.e. $\Phi(\mathbf{p})=\Phi(\pi(\mathbf{p}))$ for all profiles $\mathbf{p}=\mathbf{p}(t)$ and all λ-invariant permutations π on I; and continuity, i.e. if the sequence $\mathbf{p}_1,\mathbf{p}_2,\dots,\mathbf{p}_n,\dots$ converges to $\mathbf{p}$ almost everywhere, then $\Phi(\mathbf{p}_k)$ converges to $\Phi(\mathbf{p})$.

Let Φ be a linear continuum rule on X. Consider the characteristic function x_A with $A=[0,1/n]$, given by $x_A(t)=0$ or 1 as $t\notin A$ or $t\in A$. Anonymity, additivity and unanimity imply that $\Phi(x_A)=1/n$. From linearity and continuity it follows that Φ coincides with the Lebesgue integral on the set of simple functions, i.e. functions of the form $\lambda_1.x_{A_1}+\dots+\lambda_n.x_{A_n}$ where the $\lambda_i\in I$ and $\{A_1,\dots,A_n\}$ is a partition of I into measurable sets. Finally, since each measurable map on I is the limit of a sequence of simple functions, continuity implies that Φ coincides with the Lebesgue integral on the set $\mathscr{L}(I,X)$ (e.g. Edwards 1965, Theorem 4.8.2–4).

As I contains countable infinite sets, the continuum case extends the countable infinite population framework. However, the next argument shows that the continuum I is unsuited to deal with infinite horizon problems, where the index in payoff or consumption streams is a time index running over the ordered set $\mathbb{N}_0$. Embedding an infinite horizon problem in the continuum framework requires an order preserving embedding of the set $\mathbb{N}_0$ in I. Mapping infinite profiles on appropriate simple maps embeds I^∞ into the set $\mathscr{L}(I,X)$. But, the topology on $\mathscr{L}(I,X)$ restricted to the set of simple maps coincides with the Tychonov or the standard product topology on I^∞, and does not allow for "weakly" anonymous rules (see Efimov and Koshevoy 1994; Lauwers 1993). We conclude as follows: if continuity and anonymity are involved, the [0, 1]-continuum population framework is not the instrument to study infinite horizon problems.

References

Aczél J (1966) Lectures on functional equations and their applications. Academic Press, New York

Aumann RJ (1964) Markets with a continuum of traders. Econometrica 32: 39–50

Campbell D (1985) Impossibility theorems and infinite horizon planning. Soc Choice Welfare 2: 283–293

Candeal JC, Induráin E, Uriarte JR (1992) Some issues related to the topological aggregation of preferences. Soc Choice Welfare 9: 213–227

Candeal JC, Induráin E, Uriarte JR (1993a) Some issues related to the topological aggregation of preferences: addendum. Soc Choice Welfare 14: 359–361

Candeal JC, Induráin E (1993b) Aggregation of preferences: an algebraic overview. Preprint Universidad Publica de Navarra, Pamplona, Spain

Candeal JC, Chichilnisky G, Induráin E (1993c) Topological aggregation of preferences: the case of a continuum of agents. Preprint, Universidad Publica de Navarra, Pamplona, Spain

Chichilnisky G (1980) Social choice and the topology of spaces of preferences. Adv Math 37: 165–176

Chichilnisky G (1982) The topological equivalence of the Pareto condition and the existence of a dictator. J Math Econ 9: 223–233

Chichilnisky G (1983) Social choice and game theory: Recent results with a topological approach. In: Pattanaik PK, Salles M (eds) Social choice and welfare. North-Holland, Amsterdam

Chichilnisky G, Heal G (1979) Social choice with infinite populations. Construction of a social choice rule and impossibility results. Columbia University Economics Discussion Paper

Chichilnisky G, Heal G (1983) Necessary and sufficient conditions for the resolution of the social choice paradox. J Econ Theory 31: 68–87

Diamond PA (1965) The evaluation of infinite utility streams. Econometrica 33: 170–177

Edwards RE (1965) Functional analysis, theory and applications. Holt. Reinhard and Winston: New York, London

Efimov B, Koshevoy G (1994) The topological approach to social choice with infinite populations. J Math Soc Sci 27: 145–157

Güth W, Leininger W, Stephan G (1991) On supergames and folk theorems: a conceptual discussion. In: Selten R (ed) Game equilibrium models II. Springer, Berlin Heidelberg New York

Jameson GJO (1974) Topology and normed spaces. Chapman & Hall, London

Kirman AP, Sondermann D (1972) Arrow's theorem, many agents and invisible dictators. J Econ Theory 5: 267–277

Lauwers L (1993) Continuity and equity with infinite horizons. Soc Choice Welfare 14: 345–356

Lauwers L (1994) Rawlsian equity and generalised utilitarianism with an infinite population. Econ Theory (in press)

Lehrer E (1989) Lower equilibrium payoffs in two-player repeated games with non-observable actions. Int J Game Theory 18: 57–89

Meyer PA (1973) Limites Mediales, d'apres Mokobodzki. In: Meyer PA (ed) Seminaire de Probabilité VII, Lecture Notes in Mathematics 321, Springer, Berlin New York

Peressini A (1967) Ordered topological vector spaces. Harper and Row, New York

Schmitz N (1977) A further note on Arrow's impossibility theorem. J Math Econ 4: 189–196

Sen A (1986) Social choice theory. In: Arrow KJ, Intrilligator MD (eds) Handbook of Mathematical Economics III. North-Holland, Amsterdam

Weiss EA Jr (1981) Finitely additive exchange economies. J Math Econ 8: 221–240

Yoshida K, Hewitt E (1952) Finitely additive measures. Trans Am Math Society 72: 46–66

Zhou L (1991) Impossibility of strategy-proof mechanisms in economies with pure public goods. Rev Econ Stud 58: 107–119

Topological aggregation of preferences: the case of a continuum of agents

J.C. Candeal[1], G. Chichilnisky[2], E. Induráin[3]

[1] Departamento de Análisis Económico, Universidad de Zaragoza, Doctor Cerrada 1 y 3, E-50005 Zaragoza, Spain
[2] Department of Economics, Columbia University, 405 Law Memorial Library, New York, NY 10027, USA
[3] Departmento de Matemática e Informática, Universidad Pública de Navarra, Campus Arrosadía s.n., E-31006 Pamplona, Spain

Received: 30 November 1994/Accepted: 22 April 1996

Abstract. This paper studies the topological approach to social choice theory initiated by G. Chichilnisky (1980), extending it to the case of a continuum of agents. The social choice rules are continuous anonymous maps defined on preference spaces which respect unanimity. We establish that a social choice rule exists for a *continuum* of agents if and only if the space of preferences is contractible. We provide also a topological characterization of such rules as generalized means or mathematical expectations of individual preferences.

1. Introduction

A classical problem of social organization is how to aggregate individual into social preferences. Typical examples are voting procedures; an acceptable procedure must satisfy certain properties, or "axioms". Arrow (1951) required that the map from individual to social preferences, the "social choice rule", should be: (i) non-dictatorial, so that the outcome is not decided solely by one individual, (ii) independent from irrelevant alternatives, and (iii) Pareto, namely it should respect the unanimous wishes of the individuals between any two choices. Even though these axioms seem reasonable, without restrictions on individual preferences no such social choice rule exists. This is Arrow's impossibility theorem.

It seems natural to consider other formulations of social choice which could lead to the existence of social choice rules. Among them, the topological approach introduced in Chichilnisky (1980, 1982) deals with a topological space of preferences and requires three axioms: the aggregation rules from spaces of individual to social preferences must be (i) continuous, (ii) anonymous, i.e. invariant under permutation of the individuals, and (iii) respect unanimity, so that if everybody shares the same preference, so does society. Chichilnisky (1980, 1982) showed that these axioms also lead to an impossibility theorem when preferences are unrestricted. However, by considering restricted spaces of preferences Chichilnisky and Heal (1983) established the following result: "*A preference space*[1] *admits a social choice rule for every finite number of individuals $n \geq 2$, if and only if it is contractible*". This result, known as the *resolution of the social choice paradox*, refers to problems with finitely many individuals. It reveals the intrinsic topological structure of the problem of social choice, because it shows that social choice rules exist if and only if the preference space is contractible, i.e. if and only if their topology is trivial.

Chichilnisky and Heal (1979) studied also a natural extension of the problem to infinitely many individuals; they constructed social choice rules for problems with countably many agents, rules which are Pareto and non-dictatorial. These were constructed by taking the social preference to be a limit of sequences of individual preferences, or appropriate extensions when such limits do not exist. A *sequential passage to the limit* of the aggregation rules for finite individuals was also considered in Candeal et al. (1992). More recently, Chichilnisky (1996) introduced a concept of non-dictatorship of the present and non-dictatorship of the future in the context of economies with infinitely many generations, and established the existence and a full characterization of continuous social choice rules satisfying continuity and these two axioms.

In this paper we consider the problem of topological aggregation of preferences for a *continuum* of individuals. A continuum of individuals reflect the idea of a very large number of participants, among which each individual has a negligible impact, an idea which is central to the theory of competitive markets. A pioneer paper in this framework is Aumann (1964), where markets with a continuum of traders were studied. The problem of social choice involving a continuum of agents has not, however, been considered in the literature before.

Using a continuum framework, we establish the main result of this paper. We establish that there exists a social choice rule for a continuum of individuals if and only if the space of preferences is contractible. This is similar to the above result by Chichilniksy and Heal's (1983) on the topological structure of the social choice problem. Furthermore we establish that the social choice rules we construct are an extension of generalized averages: indeed, they are obtained by integrating or obtaining a mathematical expectation of individual's preferences.

2. Basic concepts and previous results

Basic concepts of social choice are the set of individual preferences, denoted X, and a set of individuals[2] or voters. The space of preferences X is a Hausdorff[3] topological space. The space of profiles of preferences, namely of listings of all individual preferences, is X^n when there are n individuals, and $\prod_{i=1}^{\infty} X$

---

[1] Minimal regularity conditions are required on the space of preferences; for example it must be a parafinite CW-complex. This is a mild condition which includes all locally convex spaces such as manifolds, polyhedra, all finite simplicial complexes, and many infinite dimensional spaces such as spheres in Banach spaces

[2] There could be finite, countably many or a continuum of individuals

[3] I.e. separated

when there are countably many individuals. A social choice rule is a map from spaces of individual to social preferences, $\phi: X^n \to X$ with n individuals, and $\phi: \prod_{i=1}^{\infty} X \to X$ with countably many.

We shall assume that social choice rules respect two normative properties, anonymity and respect of unanimity. Also an assumption of continuity is required.[4]

A *n-Chichilnisky rule* (see Chichilnisky 1980, 1982) on X is a map from the product space X^n into X:

$$\phi: X^n \to X$$

$$(x_1, \ldots, x_n) \to \phi(x_1, \ldots, x_n)$$

satisfying:

(i) CONTINUITY: ϕ is a continuous map (X^n is endowed with the product topology).
(ii) ANONYMITY: $\phi(x_1, \ldots, x_n) = \phi(x_{i_1}, \ldots, x_{i_n})$ for any rearrangement of $\{1, \ldots, n\}$.
(iii) RESPECT OF UNANIMITY: $\phi(x, \ldots, x) = x$ for every $n \in X$.

Similarly, in the case of an infinite but countable number of agents, one defines a *countable* ∞*-Chichilnisky rule* on X as a map:

$$\phi: X^\infty = \prod_{i=1}^{\infty} X \to X,$$

continuous, anonymous and unanimous. The concepts of anonymity and unanimity are defined similarly to the finite n-dimensional case.

The choice of topology can be important for the existence of continuous anonymous maps; anonymity is a symmetry property and as such it poses topological restrictions. Indeed, Efimov and Koshevoy (1994), and Lauwers (1997) have proved that the *product topology* (*or Tychonoff topology*) *on the space of profiles* leads to the non-existence of continuous anonymous rules respecting unanimity for countably many individuals. However if a *uniform topology* is considered on $\prod_{i=1}^{\infty} X$, then appropriate social choice rules exist. To avoid such problems one considers a weaker concept than anonymity:

(i) A *finitely anonymous countable* ∞*-Chichilnisky rule* (*or weak countable* ∞*-Chichilniksy rule*) on X is a map $\phi: X^\infty \to X$, with ϕ continuous, unanimous and *finitely anonymous.* That is, ϕ takes the same value on any two sequences such that the second is a finite rearrangement (only finitely many terms change their positions) of the former one.

As already mentioned, Chichilnisky and Heal (1983) established that when the set of individuals is finite, the existence of aggregation rules for all $n \geq 2$ is equivalent to extreme topological simplicity, indeed topological triviality. Thus, when the space of preferences X is a parafinite cellular complex (*parafinite CW-complex*), the existence of a n-rule for every n has as a necessary and sufficient condition the contractibility of X. This result, known as the

---

[4] For a motivation of the role of continuity in Social Choice theory see Chichilnisky (1982)

resolution of the Social Choice paradox is crucial in the framework of topological aggregation of preferences. The first proof of this fact appeared in Chichilnisky and Heal (1983) (see also Chichilnisky (1991) or Candeal and Induráin 1991).

With respect to the *countable infinite* case, there are several partial results (see Chichilnisky and Heal 1979 or Candeal et al. 1992). For example, we have:

(i) The existence of a countable ∞-Chichilnisky rule on a preference space X implies the existence, for every $n \in \mathbf{N}$, of a n-Chichilnisky rule on X. Thus, when X is a parafinite CW-complex, the existence, on X, of a countable ∞-Chichilnisky rule implies the contractibility of X.

(ii) Let X be a compact space of preferences. The existence on X of a n-Chichilnisky rule for every $n \in \mathbf{N}$ implies the existence of a weak countable ∞-Chichilnisky rule on X. Consequently, if X is a parafinite CW-complex, there exists on X a weak countable ∞-Chichilnisky rule.

(iii) Let X be a compact space of preferences. Then there exists a continuous, Pareto, non-dictatorial map, for countably many individuals, Chichilnisky and Heal (1979, 1994).

3. Topological Chichilnisky rules for a continuum of agents

We seek extensions of the above results to the case in which there is a continuum of individuals. To extend the notion of topological Chichilnisky rule for the case of a *continuum* of agents, we need to adequately express a *continuum Cartesian product* of copies of the preference space X. Obviously a continuum Cartesian product of X can be represented as X^I, the product of X with itself as many times as elements in I, where I has a suitable continuum cardinality, for example I could be the unit interval. The standard definition for this product space X^I is as the space of all functions from I to X.

The next step is the selection of a natural, suitable topology for X^I. We are interested in models leading to a result parallel to the "*Chichilnisky and Heal resolution of the Social Choice paradox*"[5]. In any case, the assumptions of unanimity, anonymity and continuity must be properly defined for the continuum case.

(i) Respect of unanimity.

The unanimity property means that if all the agents have the same preference (over all choices) then the social choice rule assigns to that profile that common preference. In other words, the restriction of the rule to the diagonal of the product space X^n[6], ϕ/Δ, is exactly the identity map, i.e. $\phi/\Delta = id$. To translate this property to the context of a continuum of agents, it suffices to require that the image under the social rule of *any constant map* with values in X (which is the equivalent of the diagonal Δ in the finite product

[5] *A priori*, we know nothing about the existence of continuum rules on suitable models. Of course, we do not know, at this stage, if such existence is related to additional topological properties, e.g. contractibility, on the preference space

[6] $\Delta = \{(x_1, \ldots, x_n) \in X^n; \forall\, i, j,\, x_i = x_j\}$

space X^n) is just that constant value of the map. For this property to make sense, constant X-valued maps must be included in the functional space we consider as a model for the space of profiles.

(ii) Anonymity

With respect to anonymity, we require a sensible requirement: Suppose we have a partition of the space of individuals $[0,1]$ into a finite number of coalitions, all having the same size, and that all the members of each coalition have the same partition. Then the rule must assigns the same value to any rearrangement of those over the same partition. We emphasize the condition on the partition that all its coalitions have the same size. Since the concept of size is needed to define this form of anonymity, a functional space involving some type of *measure* will be required. By technical reasons derived from the evaluation of some integrals we will work with the closed unit interval $I = [0,1]$, or any other normalized finite measurable space equipped with the *Lebesgue measure.* Thus, by definition of anonymity, the simple functions[7] supported over sets with the same measure have to be included in our model for the set of profiles as well.

(iii) Continuity

As already mentioned, the key matter here is the continuity assumption. In view of the results by Efimov and Koshevoy (1992) and Lauwers (1997), one must consider a suitable topology. This is because the countable case is "included" in the continuum one, in the sense that a continuum of individuals certainly contains a countable number of individuals. Therefore, a rule for a continuum of individuals induces a rule for countably many. As shown by Efimov and Koshevoy, the problem has no solution with the pointwise convergence topology. This implies that on the functional space X^I, one requires a topology finer than the pointwise-convergence topology, which is equivalent to the product topology in this context.

An adequate topology is the *compact-open topology*. This is the topology which has as subbasis the set of all functions which carry a given compact set $K \subset I$ into a given open set $U \subset X$. This topology is identical with the topology of uniform convergence on compacta when X is a uniform space.[8]

Having in mind the previous discussion we formalize the problem as follows.

- The space of preferences X is a uniform Hausdorff topological space, for example a metric space, and the continuum of individuals is described by the unit interval $[0,1]$ endowed with the usual topology and the Lebesgue measure μ.
- $\mathscr{S}([0,1],X)$ denotes the set of all simple maps, i.e., maps whose range is finite. A simple map s will be represented as follows:

$$s = x_1\chi_{A_1}, \ldots, x_n\chi_{A_n}$$

[7] Simple functions are maps whose range is a finite number of values

[8] For the definition of the compact open topology see Dugundji (1966), p. 257, and Kelley (1955, p. 221, Kelley (1955), p. 229, Theorem 1, proves that the compact open topology is the uniform topology on compacta when X is a uniform space, for example, when X is a metric space

where $(x_1, \ldots, x_n) \in X^n$, $(A_i)_{i=1,\ldots,n}$ is a partition of $[0,1]$ into disjoint Lebesgue-measurable sets A_i, for $i = 1, \ldots, n$, and $s(x) = x_i \Leftrightarrow x \in A_i$.

In particular, the range of s is the subset $\{x_1, \ldots, x_n\} \subset X$.

- The space of profiles, or list of individual preferences is the minimal closed space which contains the simple functions under the uniform convergence, i.e. it is the space of functions $f: [0,1] \to X$ for which there exists a sequence $(f_n)_{n \in \mathbf{N}}$ of simple maps uniformly converging of f μ-almost everywhere.[9] The space of preferences is denoted $\mathscr{L}_\infty([0,1], X)$. We endow $\mathscr{L}_\infty([0,1], X)$ with the relative compact-open topology μ-a.e.

Recall that the sets of the form

$$S(K, U) = \{f: [0,1] \to X;\ f(t) \in U,\ t \in C, \text{ where}$$
$$K \text{ is any compact subset of } [0,1], \text{ and } U \text{ any open subset of } X\},$$

constitute a subbase for this topology.

Now we define the concept of a continuum Chichilnisky rule:

- A *continuum Chichilnisky rule* on X is a map

$$\phi: \mathscr{L}_\infty([0,1], X) \to X$$

satisfying the following conditions:

(i) CONTINUITY: ϕ is continuous.
(ii) ANONYMITY: For any finite partition $(E_i)_{i=1,\ldots,n}$ of $[0,1]$ with $\mu(E_i) = 1/n$ $(i = 1, \ldots, n)$, any n-tuple $(x_1, \ldots, x_n) \in X^n$, and any perturbation of the set $\{1, \ldots, n\}$, denoted $\sigma \in S(n)$:

$$\phi(x_1\chi_{E_1}, \ldots, x_n\chi_{E_n}) = \phi(x_{\sigma(1)}\chi_{E_1}, \ldots, x_{\sigma(n)}\chi_{E_n}),$$

(iii) Respect of UNANIMITY: For every $x \in X$:

$$\phi(x\chi_{[0,1]}) = x.$$

A first result states the relationship between continuum and finite Chichilnisky rules:

Proposition 1. *The existence of a continuum Chichilnisky rule $\phi: \mathscr{L}_\infty([0,1], X) \to X$ implies the existence, for every $n \in \mathbf{N}$.[10] of a (finite) n-Chichilnisky rule on the space of preferences X, $\phi^n: X^n \to X$.*

Proof. Let $\phi: \mathscr{L}_\infty([0,1], X) \to X$ be a continuum Chichilnisky rule.

Given $n \in \mathbf{N}$, let us define the map

$$\phi_n: X^n \to X$$

by

$$\phi_n(x_1, \ldots, x_n) = \phi(x_1\chi_{[0,1/n]}, x_2\chi_{[1/n,2/n)}, \ldots, x_n\chi_{[(n-1)/n,1]}) \in X.$$

[9] Except on a set of measure zero, denoted also μ – a.e.
[10] N is the set of integers

Clearly, ϕ_n is well defined and it is anonymous and unanimous. To see that ϕ_n is continuous, it is enough to verify the continuity of the map $\phi_n^*: X^n \to \mathscr{L}_\infty([0,1], X)$ defined as

$$(x_1, \dots, x_n) \mapsto s_n = (x_1\chi_{[0,1/n]}, x_2\chi_{[1/n,2/n)}, \dots, x_n\chi_{[(n-1)/n,1]}) \in \mathscr{L}_\infty([0,1], X),$$

because ϕ_n is the composition of ϕ_n^* with ϕ, i.e. $\phi_n = \phi \circ \phi_n^*$, and ϕ is continuous. Denote by $I_i = [(i-1)/n, i/n)$, $1 \le i < n$. Notice that the preimage $(\phi_n^*)^{-1}(S(C,U))$ of a subbasic subset equals $V_1 \times \cdots \times V_n$, with $V_i = U$ if C meets I_1 μ-a.e. and $V_i = X$ otherwise, and these sets are open in the product topology of X^n. Therefore ϕ_n^*, and hence ϕ_n, is continuous. □

The following results are obtained for a wide class of preference spaces X called CW-complexes; these include euclidean spaces, all manifolds and polyhedra, and are constructed by pasting up properly a number of simple spaces called "cells". CW complexes contain a finite number of cells; examples are spheres, tori, balls, and cubes. Parafinite CW complexes may be infinite dimensional, as they may contain an infinite number of cells, although they may contain only a finite number of cells in each dimension.

Corollary 2. *Let the space of preferences X be a parafinite CW-complex. If there is a continuum Chichilnisky rule on X, then X is contractible.*

Proof. This is a direct consequence of Proposition 1 and the *Chichilnisky and Heal resolution of the social choice paradox* (see Chichilnisky and Heal 1983; Chichilnisky 1991, or Candeal and Induráin 1991). □

The next result extends Corollary 2 to show that the contractibility of X is not only necessary but is also sufficient for the existence of social choice rules.

Theorem 3. *Let the space of preferences X be a CW-complex. Then X admits a continuum Chichilnisky rule $\phi: \mathscr{L}_\infty([0,1], X) \to X$ if and only if X is a contractible space.*

Proof. Let X be a contractible CW-complex. Let $K(X)$ be the closed convex hull of X. Then $(X, K(X))$ is a cellular pair.[11] $K(X)$ is obviously contractible, since it is convex. Thus the inclusion mapping i from X into $K(X)$ is a weak homotopy equivalence.[12] Hence, it is an homotopy equivalence.[13] Therefore X is a deformation retract of $K(X)$.[14] By Whitney's theorem,[15] we may consider $K(X)$ a topological subspace of some Euclidean space $\mathbf{R}^k$, for k big enough so that in particular X is a uniform space; let "+" denote the standard addition within $\mathbf{R}^k$. Define now the map

$$\theta: \mathscr{L}_\infty([0,1], K(X)) \to K(X)$$

$$f \to \theta(f) = \int_{[0,1]} f \, d\mu.$$

[11] Rohlin and Fuchs (1981), p. 118
[12] Rohlin and Fuchs (1981), p. 445
[13] Rohlin and Fuchs (1981), p. 446
[14] Rohlin and Fuchs (1981), p. 119
[15] See, for instance, Bröcker and Jänich (1973) Satz 7.7

We first check that this map is *well defined.* Notice that any simple map $s = x_1\chi_{A_1}, \dots, x_n\chi_{A_n}$ is, obviously, integrable; indeed its integral is $x_1\mu(A_1) + \cdots + x_n\mu(A_n) \in K(X)$.

Now if f is a uniform limit of simple functions, it is μ – *a.e.* bounded, and therefore it is a Lebesgue-integrable as a map from $[0,1]$ to $\mathbf{R}^k$; in fact by the Lebesgue bounded convergence theorem, its integral is a limit in $\mathbf{R}^k$ of elements that belong to $K(X)$. But $K(X)$ is closed in $\mathbf{R}^k$. Therefore $\theta(f)$ must lie in $K(X)$.

Now since $(X, K(X))$ is a cellular pair, there exists a continuous *retraction*[16] from $K(X)$ into X, denoted

$$\rho : K(X) \to X. \tag{1}$$

Consider therefore the following composition of maps $\phi = j \circ \theta \circ \rho$, which maps $\mathscr{L}_\infty([0,1], X)$ into X:

$$\mathscr{L}_\infty([0,1], X) \xrightarrow{j} \mathscr{L}_\infty([0,1], K(X)) \xrightarrow{\theta} K(X) \xrightarrow{\rho} X$$

where j is the inclusion map. *This composition is a continuum Chichilnisky rule defined on X.* □

On parafinite contractible spaces, Chichilnisky topological n-rules are, essentially, retractions of convex means (see, e.g., Chichilnisky and Heal 1983). Here, working with the *continuum* case, we have obtained a parallel result:

Corollary 4. *Any continuous, anonymous social choice rule* $\phi : \mathscr{L}_\infty([0,1], X) \to X$ *respecting unanimity is homotopic to a mathematical expectation of individual preferences.*

Proof. By the previous theorem a map ϕ with the desired properties exists if and only if X is contractible. Note that the map $\phi = j \circ \theta \circ \rho$ constructed above is a mathematical expectation i.e. it has the desired property. Furthermore, when X is contractible, any other map $\phi : \mathscr{L}_\infty([0,1], X) \to X$ is homotopic to $j \circ \theta \circ \rho$. This completes the proof. □

Theorem 3 has been stated for *finite* CW-complexes (a finite union of cells of finite dimension). However, Chichilnisky and Heal's resolution of the Social Choice paradox was stated for the more general case of closed *parafinite CW-complexes,* consisting of an appropriate union of cells which could include infinitely many cells, but such that the number of cells of each finite dimension is finite. It is possible to strengthen Theorem 3 to show that the analogous result is valid for the parafinite case.

Theorem 5. *Let the space of preferences be a parafinite CW complex. Then X admits a continuum Chichilnisky rule* $\phi : \mathscr{L}_\infty([0,1], X)$ *if and only if X is a contractible space.*

Proof. The strategy of this proof is as follows. First we define a social choice map on simple profiles in $\mathscr{L}_\infty([0,1], X) \to X$, i.e. profiles defined by simple maps, and then show that this rule can be extended to all profiles $f \in \mathscr{L}_\infty([0,1], X)$,

---

[16] A retraction from a space Y into a subspace $X \subset Y$ is a continuous map $r : Y \to X : \forall x \in X, r(x) = x$

since each f can be approximated uniformly by simple maps. The definition of the social choice rule on simple map is easy to achieve: by definition, the image of a simple map is contained in the union of all cells of X up to some finite dimension, say n; this union is called the n-skeleton of X, and is denoted X_n. The integral of each simple map is therefore contained in the convex hull of X_n, denoted $K(X_n)$. The integral is shown to be well defined and it is a continuous, Pareto and non-dictatorial map on simple profiles. Next we show that the limit of these integrals defines the integral of f; the map assigning to f its integral, which is the natural extension of the rule defined in Theorem 3, is the desired social choice rule and completes the proof.

Let s be a simple map in $\mathscr{L}_\infty([0,1],X)$. The image of s consists of finitely many points and is contained in some skeleton of X, say the *n-th* skeleton of X, X_n, so that its integral $\int_{[0,1]} s\,d\mu$ is contained in the convex hull of X_n, $K(X_n)$. Compose this integral with the retraction map ρ_n from $K(X_n)$ into X_n defined in (1) above. This defines the value for the integral of s in X, denoted $\phi(s_n) \in X$

$$\phi_n(s) = \rho_n \circ \int_{[0,1]} s\,d\mu.$$

It is immediate to check that the above construction can be made independent from the choice of n-skeletons, by choosing inductively the retraction maps ρ_n from $K(X_n)$ onto X_n so that they agree on subskeletons, i.e. so that

$$X_n \subset X_q \;\Rightarrow\; \rho_n : K(X_n) \to X_n = \rho_{q/X_n} : K(X_n) \to X_n,$$

where ρ_{q/X_n} is the restriction of the map ρ_q to the subset $X_n \subset X_q$. One therefore has defined inductively a map $\rho: \cup_n K(X_n) \to X$, such that $\rho/K(X_n) = \rho_n$.

The composition map $\phi(s) = \rho \circ \int_{[0,1]} s\,d\mu$ defines a social choice map on the subspace of simple maps of $\mathscr{L}_\infty([0,1],X)$. The map ϕ is similar to the social choice map defined in Theorem 3; in particular it is continuous, non-dictatorial and Pareto. The next step is to extend ϕ to *all* maps in $\mathscr{L}_\infty([0,1],X)$.

Recall that any f in $\mathscr{L}_\infty([0,1],X)$ is the uniform limit of simple maps $(s_n)_{n\in N}$. The integrals $\int_{[0,1]} s_n\,d\mu$ in fact define a Cauchy sequence in R^∞, i.e. given a convex neighborhood U of the origin in the space of all sequences with the product topology, R^∞, there exists $n_0 \in N$ such that

$$\int_{[0,1]} (s_n - s_m)\,d\mu \in U$$

for every n, $m > n_0$. Now $\int_{[0,1]} s_n\,d\mu \in K(X)$ for every n and $K(X)$ is closed, so that

$$\int_{[0,1]} f\,d\mu = \lim_{n\to\infty} \int_{[0,1]} s_n\,d\mu \in K(X).$$

It is routine to check that this definition is consistent, i.e. it does not depend on the sequence of simple maps (s_n) chosen. Thus the map $\phi: \mathscr{L}_\infty([0,1],X) \to X$ given by $\phi(f) = \int_{[0,1]} f\,d\mu$ is the desired social choice rule. □

Remark 1. The set $\mathscr{L}_\infty([0,1],K(X))$ used in the previous discussion may be substituted by a larger set, that we shall denote $\mathscr{L}_1([0,1],K(X))$ in the same way that $\mathscr{L}_\infty([0,1])$ is contained in $\mathscr{L}_1([0,1])$ and following the steps of the construction of the Lebesgue integral on $[0,1]$.

Thus, we say that a map $f\colon[0,1]\to K(X)$ is *integrable*, if embedding $K(X)$ in $\mathbf{R}^k$, via Whitney's theorem, the map $f^*\colon[0,1]\to\mathbf{R}^k$ is Lebesgue integrable, f^* equal to the composition $i\circ f$, with i the embedding map $i\colon K(X)\to\mathbf{R}^k$ given by Whitney's theorem.

If f is integrable, its integral, denoted by $\int_{[0,1]} f$, will be the following element of $K(X)$:

$$\int_{[0,1]} f = i^{-1}\left(\int_{[0,1]} (i\circ f)\,\mathrm{d}\mu\right).$$

Now define the set $\mathscr{L}_1([0,1],K(X))$ as follows: $\mathscr{L}_1([0,1],K(X))=\{f\colon [0,1]\to K(X);$ is integrable$\}$, and observe that Theorem 3 and its Corollary can be generalized using $\mathscr{L}_1([0,1],K(X))$ instead of $\mathscr{L}_\infty([0,1],K(X))$. □

Remark 2. Recently Lauwers and van Liedekerke (1993) have proved an analogue of the Chichilnisky-Heal resolution of the social choice paradox in the context of countably many agents. Their result is phrased in terms of a new concept of anonymity, that they call *bounded anonymous infinite rules.*

References

1. Arrow KJ (1951) Social choice and individual values. Wiley, New York
2. Aumann RJ (1964) Markets with a continuum of traders. Econometrica 32: 39–50
3. Bröcker T, Jänich K (1973) Einführung in die Differentialtopologie, Springer, Berlin
4. Candeal JC, Induráin E (1991) Aggregation of preferences: an algebraic overview. (Preprint). DTAEUZ no. 91-8, Universidad de Zaragoza, 1991
5. Candeal JC, Induráin E, Uriarte JR (1992) Some issues related to the topological aggregation of preferences. Soc Choice Welfare 9: 213–227
6. Chichilnisky G (1980) Social Choice and the topology of the spaces of preferences. Adv Math 37: 165–176
7. Chichilnisky G (1982) Social aggregation rules and continuity. Q J Econ 96: 337–352
8. Chichilnisky G (1991) Actions of symmetry groups in Social Choice (Preprint). Columbia University, New York
9. Chichilnisky G (1996) An axiomatic approach to sustainable development. Soc Choice Welfare 13: 231–257
10. Chichilnisky G, Heal G (1979) Social choice with infinite populations. Construction of a Social Choice rule and impossibility results. Columbia University, Economics Department Discussion paper, New York
11. Chichilnisky G, Heal G (1983) Necessary and sufficient conditions for a resolution of the Social Choice paradox. J Econ Theory 31: 68–87
12. Dugundji J (1966) Topology. Allyn and Bacon, Boston
13. Efimov BA, Koshevoy GA (1994) The topological approach to social choice with infinite populations. Math Soc Sciences 27(2): 145–157
14. Kelley J (1955) General Topology. D. Van Nostrand, Inc. Princeton, NJ New York
15. Lauwers L (1997) A note on weak ∞-rules. Soc Choice Welfare 14: 357–358

16. Lauwers L, van Liedekerke L (1993) Monotonic Chichilnisky rules with infinite population. Working Paper, Monitoraat E.T.E.W. Katholieke Universiteit Leuven, Leuven, Belgium
17. Rohlin VA, Fuchs DB (1981) Premier cours de Topologie, Chapitres Géométriques. Mir, Moscow French translation: the Russian original was published in 1977. Mir Publishers, Moscow; English translation (1984) The beginner's course in Topology. Springer, Berlin Heidelberg, New York

Continuity and equity with infinite horizons

Luc Lauwers

Monitoraat ETEW, KU Leuven, Dekenstraat 2, B-3000 Leuven, Belgium

Received: 15 April 1993/Accepted: 22 April 1996

Abstract. In an infinite dimensional space, e.g. the set of infinite utility streams, there is no natural topology and the content of continuity is manipulable. Different desirable properties induce different topologies. We consider three properties: effectiveness. l_1-summability and equity. In view of effectivity, the product topology is the most favourable one. The strict topology is the largest topology for which all the continuous linear maps are l_1-summable. However, both topologies are myopic and conflict with the principle of equity. In case equity is desirable, the sup topology comes forward.

1. Introduction

This overview of different topologies is motivated by a recent debate on the extension to the infinite population case of the topological approach to social choice theory as initiated by Chichilnisky (1980, 1982). The importance of this approach was revealed in Chichilnisky and Heal (1983), where the resolution of the social choice paradox was proved: a Hausdorff topological space of preferences allows for continuous, unanimous and anonymous maps from the finite cartesian products into itself if and only if it is contractible. When looking for an infinite version of the continuity demand, difficulties arised in the choice of an appropriate topology.[1]

A parallel debate was held in the literature initiated by Koopmans (1960, 1972) and dealing with the representation of preferences over time. In this framework, intertemporal choice is reflected in the evaluation of infinite utility streams, i.e. sequences the coordinates of which represent the one period utility levels of some alternative. This tradition also uses continuity as an

The author is grateful to Geoffrey Heal for many valuable comments.

[1] Chichilnisky and Heal (1979), Candeal et al. (1992), Lauwers (1992, 1993b) and Efimov and Koshevoy (1992) deal with infinite Chichilnisky rules.

appealing axiom. Again, problems emerged in specifying the topology upon the set l_∞ of infinite utility streams. To wit, since the continuity demand becomes weaker as a topology expands, Svensson (1980) was able to construct a topology on the set l_∞ large enough to allow for a continuous, strongly monotonic and equitable evaluation. A well established impossibility result of Diamond (1965) seemed to be resolved. As noticed by Campbell (1985), the construction of the Svensson topology revealed that continuity can be imposed with impunity. This observation turns the continuity axiom into an empty axiom, and results in an explicit refusal to equip the set l_∞ with a topology (e.g. Epstein 1986a, b and Harvey 1986).

This note reconsiders the use of a topology on l_∞ and offers arguments in favour of a continuity demand. Continuity has an attractive interpretation: if an evaluation ranks a stream above a second one, continuity implies that utility streams close to the first are also ranked above streams close to the second. Second, a continuous map reaches extreme values on compact sets. Hence, in optimization problems continuity implies effectiveness (i.e. the existence of optimizers) in case the feasible sets are compact. Third, Brown and Lewis (1981) motivate the use of a topology as a behavioural assumption reflecting the myopic behaviour of economic agents.

Brown and Lewis (1981) observe that the strict topology $\mathscr{T}_m$ is the largest Hausdorff locally convex topology for which continuous linear maps on l_∞ are represented as l_1-summable sequences. When looking for effectiveness. Tychonov's theorem implies that the product topology $\mathscr{T}_p$ is the most favourable candidate. Indeed, the feasible sets as they appear in the literature are compact with respect to this topology.

However, both $\mathscr{T}_m$ and $\mathscr{T}_p$ are unable to deal with equity. Hence, if equity is a desirable property, these topologies become unattractive. And, at this point the link with Chichilnisky aggregation reappears: Proposition 1 implies that an equitable preference order on l_∞ in the Koopmans framework is representable by an infinite Chichilnisky rule on the set of real numbers. Such rules impose besides continuity an anonymity condition which is equivalent with equity. Combining both axioms puts severe constraints on the type of topology.

Theorem 1 shows that only topologies larger than the uniform or sup topology $\mathscr{T}_u$ are able to deal with equity. Theorem 2 uses the relationship between finite and infinite Chichilnisky rules to complete the characterization of $\mathscr{T}_u$. This characterization removes a lot of the arbitrariness from the choice of a topology in an infinite setting: continuity with respect to the sup topology is the most appropriate infinite version of the continuity demand.

The next section provides notation. Section 3 lists five different topologies on l_∞: the Tychonov or product topology, the strict, the Campbell, the sup, and the Svensson topology. Proposition 1 provides a condition upon a topology that guarantees the representability of a continuous preference order on l_∞. Section 4 deals with the Tychonov topology, and Sect. 5 with the strict topology. Both sections mainly collect existing results. The Campbell topology (Campbell 1985) appears to be an analogue of the strict topology. Continuity with respect to these topologies and equity are incompatible. Section 6 studies the sup topology and proves Theorems 1 and 2. Both theorems are new, they justify the use of the sup topology in infinite frameworks. Section 7 studies the Svensson topology. The attempt to reconcile equity, strong monotonicity and continuity by means of the von Weiszäcker criterion is analysed. Section 8 summarizes the results of the paper.

2. Notation

Let $\mathbb{R}$ be the set of real numbers and let $\mathbb{R}^\infty$ be the infinite cartesian product with $\infty = |\mathbb{N}|$. Elements of $\mathbb{R}^\infty$ or $\mathbb{R}^n$ are denoted by bold faced letters, e.g. $\mathbf{u} = (u_1, u_2, \ldots, u_t, \ldots)$. The subset $\{\mathbf{u} | \exists b \in \mathbb{R}^+$ such that $\forall k \in \mathcal{N}_0$ we have $|u_k| \leq b\}$ of bounded vectors is denoted by l_∞. An element $\mathbf{u}$ of l_∞ is called an infinite utility stream with u_t the one period utility level at time t. We write (u) for the constant stream $(u, u, \ldots, u, \ldots) \in l_\infty$. The stream (0) is denoted by $\mathbf{o}$, and $\mathbf{e}_k$ stands for the stream $(0, \ldots, 1, 0, \ldots, 0, \ldots)$ with the 1 at the kth place. Let $\mathbf{y} = (y_1, y_2, \ldots, y_n) \in \mathbb{R}^n$ and $\mathbf{u} = (u_1, u_2, \ldots, u_s, \ldots, u_t, \ldots) \in l_\infty$. Then $(\mathbf{y}, \mathbf{u})$ stands for the stream $(y_1, \ldots, y_n, u_1, u_2, \ldots, u_s, \ldots)$ and ${}_{\text{rep}}\mathbf{y}$ denotes the stream $(y_1, y_2, \ldots, y_n; y_1, y_2, \ldots, y_n; \ldots; y_1, y_2, \ldots, y_n; \ldots)$ which repeats $\mathbf{y}$ indefinitely. The truncated part $(u_1, \ldots, u_n)$ is denoted by $\mathbf{u}^n$. Finally, the infinite utility stream $(u_1, u_2, \ldots, u_t, \ldots, u_s, \ldots)$ obtained from $\mathbf{u}$ after exchanging u_t and u_s is denoted $\mathbf{u}^{t\sim s}$.

The partial order $\geq$, defined by $\mathbf{u} \geq \mathbf{v}$ if for all $k \in \mathbb{N}_0$ we have $u_k \geq v_k$, turns the space l_∞ into a vector lattice. A map $l_\infty : \mathcal{L} \to \mathbb{R}$ is said to be order bounded if sets of the form $\{\mathbf{u} | \mathbf{v}_1 \geq \mathbf{u} \geq \mathbf{v}_2\} \subseteq l_\infty$ with $\mathbf{v}_1, \mathbf{v}_2 \in l_\infty$ are mapped into closed intervals of $\mathbb{R}$. A map $F: l_\infty \to \mathbb{R}$ is said to be monotonic in case (*i*) $\mathbf{u} \geq \mathbf{v}$ implies $F(\mathbf{u}) \geq F(\mathbf{v})$ and (*ii*) $u > v$ implies $F((u)) > F((v))$.

A preference order $\succsim$ on l_∞ is a complete, reflexive and transitive binary relation in l_∞. Application of $\succsim$ on the pair $\mathbf{u}, \mathbf{v}$ of streams is denoted by $\mathbf{u} \succsim \mathbf{v}$ and means that $\mathbf{u}$ is preferred to $\mathbf{v}$. The strict order $\succ$ and the indifference relation $\sim$ are defined as usual. Let $\succsim$ be a preference order on l_∞. Then, $\succsim$ is said to be monotonic in case (*i*) $\mathbf{u} \geq \mathbf{v}$ implies $\mathbf{u} \succsim \mathbf{v}$ and (*ii*) $u > v$ implies $(u) \succ (v)$. The order $\succsim$ is said to be strongly monotonic if $\mathbf{u} \geq \mathbf{v}$ and $\mathbf{u} \neq \mathbf{v}$ implies $\mathbf{u} \succ \mathbf{v}$. Finally, the order $\succsim$ is said to satisfy Diamond's equity principle if for all utility streams $\mathbf{u}$ and all $t, s \in \mathbb{N}_0$, we have $\mathbf{u} \sim \mathbf{u}^{t\sim s}$ (Diamond 1965).

Let l_∞ be equipped with a topology $\mathcal{T}$. A preference order $\succsim$ on l_∞ is said to be continuous with respect to $\mathcal{T}$ if all upper and lower contour sets are closed with respect to $\mathcal{T}$. Equivalently, $\succsim$ is continuous if any convergent sequence of points, all of which are at least (resp. at most) as good as a given point has a limit which also is at least (resp. at most) as good as the given point.

Denote the Euclidean topology on $\mathbb{R}^k$ by $\mathcal{N}$, for $k \in \mathbb{N}_0$. For $k = 1$, this topology is induced by the Euclidean metric d_E given by $d_E(x, y) = |x - y|$. Since $\mathcal{N}$ is experienced as the natural topology on $\mathbb{R}^k$, we only consider topologies $\mathcal{T}$ on l_∞ that extend $\mathcal{N}$, i.e. topologies $\mathcal{T}$ for which the relative topology on $\mathbb{R}^k$ coincides with $\mathcal{N}$, for all $k \in \mathbb{N}_0$.

3. Five topologies on l_∞

In this section we list five topologies on l_∞ which have appeared in the literature. Since all these topologies are induced by a metric, it is sufficient to define the distance functions.

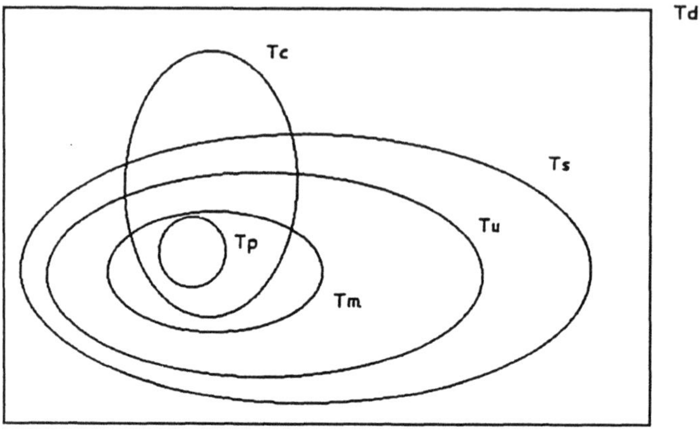

Fig. 1

Definition 1. The following real valued maps on the set $l_\infty \times l_\infty$ are distance functions. Let $\mathbf{u}, \mathbf{v} \in l_\infty$.

(i) $d_p(\mathbf{u}, \mathbf{v}) = \sum_{k=1}^{\infty} 2^{-k} |u_k - v_k|$,

(ii) $d_u(\mathbf{u}, \mathbf{v}) = \text{supremum } \{|u_k - v_k| \, | k \in \mathbb{N}_0\}$.

(iii) $d_m(\mathbf{u}, \mathbf{v}) = \text{supremum } \{|u_k - v_k|/k \, | k \in \mathbb{N}_0\}$,

(iv) $d_c(\mathbf{u}, \mathbf{v}) = \text{supremum } \{\delta(u_k, v_k)/k \, | k \in \mathbb{N}_0\}$, where

$$\delta(u, v) = \begin{cases} 0 & \text{if } u = v \\ 1 & \text{if } u \neq v \end{cases}, \text{ and}$$

(v) $d_s(\mathbf{u}, \mathbf{v}) = \text{minimum } \{1; \sum_{k=1}^{\infty} |u_k - v_k|\}$.

The distance function d_p generates the product topology $\mathscr{T}_p$:d_u the sup or uniform topology $\mathscr{T}_u$: and d_m the strict or myopic topology $\mathscr{T}_m$.[2] The distance function d_c is constructed by Campbell (1985). The corresponding topology $\mathscr{T}_c$ is called the Campbell topology. Finally, d_s generates the Svensson topology $\mathscr{T}_s$ (Svensson 1980). A property with respect to a topology $\mathscr{S}$ will be called a $\mathscr{S}$-property. Observe that all topologies except $\mathscr{T}_c$ coincide with the Euclidean topology when restricted to finite dimensions. The following inclusions appear (Brown and Lewis 1981; Campbell 1985; Shinotsuka 1994):

$$\mathscr{T}_p \subset \mathscr{T}_m \subset \mathscr{T}_u \subset \mathscr{T}_s \quad \text{and} \quad \mathscr{T}_p \subset \mathscr{T}_c,$$

Also, $\mathscr{T}_c$ overlaps with $\mathscr{T}_s$ as indicated by Fig. 1. $\mathscr{T}_d$ stands for the discrete topology: with respect to $\mathscr{T}_d$ all subsets of l_∞ are open sets.

The next proposition derives conditions on the topology for the representability by a real map of a continuous preference order. The following condition appears to be sufficient.

Definition 2. A topology $\mathscr{S}$ on l_∞ is said to satisfy the diagonal convergence condition if the convergence of a monotonic sequence $r_1, r_2, \ldots, r_n, \ldots$ of real

[2] The strict topology $\mathscr{T}_m$ defined in this paper is slightly smaller than the strict topology defined in Conway (1966). With respect to the properties dealt with in this paper both topologies behave similar (see also Shinotsuka 1994).

numbers implies the $\mathscr{S}$-convergence of the sequence $(r_1), (r_2), \ldots, (r_n), \ldots$ of constant utility streams to $(\lim_n r_n)$.

Topologies $\mathscr{S}$ smaller than $\mathscr{T}_u$ satisfy this condition: the Campbell topology $\mathscr{T}_c$ and the Svensson topology $\mathscr{T}_s$ don't. Proposition 1 implies that the search for a utility stream, maximal with respect to a $\mathscr{T}_u$-continuous preference order reduces to the maximization of a real valued map.

Proposition 1. *Let* $\mathbb{R}$ *be equipped with the standard topology* $\mathscr{N}$ *and* l_∞ *with a topology* $\mathscr{S}$ *satisfying the diagonal convergence condition. Then, a continuous and monotonic preference order* $\succsim$ *upon* l_∞*, is representable by a continuous map* $V: l_\infty \to \mathbb{R}$. *In addition, y is order bounded, monotonic and satisfies* $V((u)) = u$ *for all* $u \in \mathbb{R}$.

Proof.[3] For a stream $\mathbf{u} \in l_\infty$, define the set $C^u = \{c \in \mathbb{R} \mid (c) \succsim \mathbf{u}\}$ and $C_u = \{c \in \mathbb{R} \mid \mathbf{u} \succsim (c)\}$. Since $\mathbf{u}$ is a bounded vector, monotonicity implies that both C^u and C_u are nonempty sets. Also, $\inf C^u \geq \sup C_u$. Completeness and monotonicity of $\succsim$ imply that $u^* = \inf C^u = \sup C_u$. Since $\mathscr{S}$ satisfies the diagonal convergence condition, a sequence u_k of real numbers in C_u converging to u^* induces an $\mathscr{S}$-convergent sequence (u_k) of constant utility streams with limit equal to (u^*). Continuity of $\succsim$ implies that $(u^*) \succsim \mathbf{u}$. Analogously, one shows $\mathbf{u} \succsim (u^*)$. Hence, for a stream $\mathbf{u} \in l_\infty$, there is a unique $u^* \in \mathbb{R}$ such that $\mathbf{u} \sim (u^*)$. Define $V(\mathbf{u}) = u^*$. As $\mathbb{R}$ is equipped with the standard topology, the continuity of V follows from the continuity of $\succsim$ with respect to $\mathscr{S}$.

The property "$V((u)) = u$ for all $u \in \mathbb{R}$" formally coincides with the condition "V is unanimous". In addition, the equitability of an order $\succsim$ is reflected in its representation V: the preference order $\succsim$ on l_∞ satisfies Diamond's equity principle if and only if V satisfies the condition $V(\mathbf{u}) = V(\mathbf{u}^{t \sim s})$ for all $\mathbf{u} \in l_\infty$ and $t, s \in \mathbb{N}_0$. These observations link the ranking of utility streams with the topological approach to social choice.

Recall that an infinite Chichilnisky rule on a topological space X of preferences is a continuous, unanimous, and equitable or anonymous map $F: X^\infty \to X$. In this setting, the unanimity condition "$F((x)) = x$" requires that if all individuals have identical preferences $x \in X$, the social choice rule yields the same preference x as outcome.

Conclude that an equitable and monotonic preference order on l_∞, continuous with respect to the sup topology, is represented by a monotonic infinite Chichilnisky rule on the set $\mathbb{R}$ of real numbers.

4. Compact sets and effectivity

Following Campbell (1985, p. 285), the use of a topology can be motivated in case continuity implies the existence of extreme elements on an exogenously given family B of feasible sets. From this point of view, a good candidate for a topology would be the largest topology in which every member of B is

[3] The proof of this proposition closely follows Diamond (1965, note 7) and Kannai (1970, p. 798).

compact. Indeed, imposing continuity upon the objective function guarantees the existence of optimal elements on a feasible set. In that way a topology becomes a servant. In order to determine the family of all compact sets, Tychonov's theorem is extremely relevant.

This theorem states that in case $(X, \mathscr{S})$ is a topological space and the cartesian product X^α (where α is any number) is equipped with the product or Tychonov topology $\mathscr{S}_p$, the compact sets are exactly the cartesian products of sets which are compact with respect to $\mathscr{S}$. In case $\alpha = \infty$ and $\mathscr{S}$ is induced by a distance function d_0, then the product topology $\mathscr{S}_p$ is induced by the metric d_∞ defined by

$$d_\infty(\mathbf{x}, \mathbf{y}) = \sum_{k=1}^{\infty} 2^{-k} \cdot d_0(x_k, y_k),$$

for $\mathbf{x}, \mathbf{y} \in X^\infty$ [e.g. Kelley (1955)].

When $(X, \mathscr{S})$ coincides with $(\mathbb{R}, \mathscr{N})$, the product topology $\mathscr{N}_p$ is equal to $\mathscr{T}_p$. By means of Tychonov's theorem sets of the form

$$\{\mathbf{x} \mid \mathbf{x} \geq \mathbf{0} \text{ and } \sum x_k \cdot p_k \leq 1\},$$

with $\mathbf{p}$ a price profile satisfying $p_k > 0$ for all k, and $\mathbf{x}$ a consumption path (as in Epstein 1986b) are compact with respect to $\mathscr{T}_p$. Conclude that in view of effectiveness, the product topology is very attractive.

The frequently used discounting rule V_β, defined by

$$V_\beta : \mathscr{L} \to \mathbb{R} : \mathbf{u} \mapsto \sum_{t=1}^{\infty} \beta^{t-1} . u_t .$$

with $\beta \in]0, 1[$, is continuous with respect to $\mathscr{T}_p$. Consequently, the use of V_β as an objective function in optimization problems avoids complications with respect to the existence of extremes.

5. l_1-summability and myopia

Dealing with infinite horizon choice in a Walrasian exchange economy. Brown and Lewis (1981) focus on real values linear maps on l_∞. Their results depend on the Yosida–Hewitt representation theorem. This theorem decomposes a $\mathscr{T}_u$-continuous linear map $F : l_\infty \to \mathbb{R}$ into a sum

$$F = F^p + F^s$$

where F^p is a purely finitely additive map and F^s is l_1-summable, i.e. $F^p(\mathbf{u}) = 0$, for all $\mathbf{u} \in l_\infty$ with at most a finite number of nonzero coordinates, and

$$F^s(\mathbf{u}) = \sum_{k=1}^{\infty} \alpha_k \cdot u_k, \text{ with } \mathbf{u} \in l_\infty \text{ and } \sum_{k=1}^{\infty} \alpha_k \in \mathbb{R} \text{ (Peressini 1967).}$$

Given the Walrasian framework, conditions like the marginal rate of substitution between two goods being equal to the ratio of their prices become inoperable given such a purely finitely additive function. This in mind, Brown and Lewis (1981, p. 362) look for a notion of continuity strong enough to kill these purely finitely additive parts.

The strict topology $\mathcal{T}_m$ meets this requirement. The purely finitely additive part F^p of F is continuous with respect to $\mathcal{T}_m$ if and only if F^p is identical to zero. Indeed, for a utility stream **u**, the sequence $(\mathbf{u}^1, \mathbf{0})$, $(\mathbf{u}^2, \mathbf{0}), \ldots, (\mathbf{u}^n, \mathbf{0}), \ldots$ $\mathcal{T}_m$-converges to **u**. Since $F^p(\mathbf{u}^n, \mathbf{0}) = 0$ for all n, $\mathcal{T}_m$-continuity implies that $F^p(\mathbf{u}) = 0$. Hence, a $\mathcal{T}_m$-continuous linear map F coincides with its l_1-summable part. In addition, the topology $\mathcal{T}_m$ is the largest topology having this property.

But confront this topology with Diamond's equity axiom. Lemma 1 shows that a $\mathcal{T}_m$-continuous linear map cannot be equitable. Even more, such a map exhibits eventual impatience. By eventual impatience is meant that for t large enough, the inequality $u_t > u_1$ implies that the stream $\mathbf{u}^{1\sim t}$ is strictly preferred to **u**, i.e. $F(\mathbf{u}^{1\sim t}) > F(\mathbf{u})$ (see Diamond 1965 and Fishburn and Rubinstein 1982). One prefers to have the higher utility level now rather than in the far future.

Lemma 1. *Let $F: l_\infty \to \mathbb{R}$ be a monotonic $\mathcal{T}_m$-continuous linear map. Assume that $F(\mathbf{e}_1) \neq 0$. Then, F exhibits eventual impatience.*

Proof. Clearly F coincides with its l_1-summable part. Since, F is monotonic $F(\mathbf{u}) = \sum_{k=1}^{\infty} \alpha_k \cdot u_k$ with $\alpha_k \geq 0$ and $\alpha_1 > 0$. The convergency of the series $\sum \alpha_k$ implies $\lim_{n\to\infty} \alpha_n = 0$. Hence, for t large enough we have $\alpha_t < \alpha_1$.

In the next lemma the linearity condition is dropped. Continuity with respect to a myopic topology and the equity principle are shown to be mutually exclusive.

Lemma 2. *Let $F: l_\infty \to \mathbb{R}$ be a monotonic $\mathcal{T}_m$-continuous map. Then F is not equitable.*

Proof. Since $\mathcal{T}_p \subset \mathcal{T}_m$, continuity with respect to $\mathcal{T}_p$ implies continuity with respect to $\mathcal{T}_m$. Hence, this lemma is stronger than the theorem in Lauwers (1997). Nevertheless, its proof remains valid. Denote $\mathbf{D}_k = (1, \ldots, 1, \mathbf{0})$ with the last 1 at the k'th place. Then, $\mathbf{D}_0 = \mathbf{0}$ and $\mathbf{D}_k^{k\sim n}$ converges to $\mathbf{D}_{k-1}$ as n goes to infinity. Imposing equity and continuity implies that $F(\mathbf{D}_k) = F(\mathbf{D}_{k-1})$ for all $k \in \mathcal{N}_0$. Since the sequence $\mathbf{D}_k$ $\mathcal{T}_m$-converges to the constant stream (1), F cannot satisfy the monotonicity demand. A contradiction.

The convergence of the sequence $\mathbf{D}_k$ to (1) illustrates the myopic behaviour of the strict topology. As k increases, this topology becomes unable to discriminate between a vector with a large but finite number of coordinates equal to 1 and the constant stream (1).

A look at the metric d_c, reveals that also the Campbell topology exhibits the same myopic behaviour: the sequence $\mathbf{D}_k$ $\mathcal{T}_c$-converges to (1). The metric d_c looks similar as d_m, and the topology $\mathcal{T}_c$ can be seen as the strict or myopic topology based upon the discrete topology on $\mathbb{R}$ [the discrete topology on $\mathcal{R}$ is induced by the metric δ].[4] Therefore, the incompatibility between $\mathcal{T}_c$-continuity and equity, as proved by Campbell (1985) is not surprising.

---

[4] Campbell calls $\mathcal{T}_c$ a time decentralising topology. An infinite sequence $\mathbf{u}_1, \mathbf{u}_2, \ldots, \mathbf{u}_t \ldots$ $\mathcal{T}_c$-converges to **u**, in case the vectors $\mathbf{u}_t$ and **u** coincide on initial parts the length of which increases over time t. Time decentralising means that for each point t in time, the history of the utility stream **u**, i.e. the initial part $\mathbf{u}^{t-1} = (u_1, u_2, \ldots, u_{t-1})$, cannot change any more.

6. Equity and the sup topology

In the construction of an intergenerational choice function, concern towards future generations leads to the imposition of an equity axiom. The weakest version of equity, i.e. Diamond's equity principle, coincides with the absence of eventual impatience.[5] The next theorem states that in order to guarantee the existence of a continuous, monotonic and equitable preference order on l_∞, the set l_∞ needs a topology that is larger than the uniform topology $\mathcal{T}_u$.

Theorem 1. *Let $\mathcal{S}$ be topology induced by a norm on l_∞ and let $V:l_\infty \to \mathbb{R}$ represent a continuous, monotonic and equitable preference order on l_∞. Then $\mathcal{T}_u \subseteq \mathcal{S}$.*

The intuition behind this theorem is easy to grasp. Consider the vector $\mathbf{e}_t$ which represent the tth generation. With respect to the myopic topologies $\mathcal{T}_p$, $\mathcal{T}_m$ and $\mathcal{T}_c$, the distance between $\mathbf{e}_t$ and $\mathbf{0}$ becomes smaller as t increases. Hence, a myopic topology does not treat the generations in an equal way. On the other hand, with respect to the uniform topology the distance between $\mathbf{e}_t$ and $\mathbf{0}$ is t-independent.

Proof of Theorem 1. Recall that a map is continuous if perturbations in the outcome are small as soon they are generated by sufficiently small perturbations on the domain. Let 0 be an open set of small perturbations containing $\mathbf{0}$. Note that 0 is a full set, i.e. if $\mathbf{u}_1, \mathbf{u}_2 \in 0$ and $\mathbf{u}_1 \leq \mathbf{v} \leq \mathbf{u}_2$, then $\mathbf{v} \in 0$. If $\mathcal{S}$ is able to deal with equity, then 0 has to be closed for finite permutations. Let $(\delta, \mathbf{0}) \notin 0$ be such that $(\delta', \mathbf{0}) \in 0$ as soon $-\delta < \delta' < \delta \in \mathbb{R}$. The largest full set, closed for finite permutations and not containing $(\delta, \mathbf{0})$, is of the form $]-\delta, \delta[^\infty$. The topology generated by these open sets of perturbations coincides with $\mathcal{T}_u$.

In view of Theorem 1, it is clear that only topologies larger than the uniform topology are appropriate to develop infinite Chichilnisky rules. Hence, $\mathcal{T}_u$ can be seen as a lower bound for such topologies.

The technique of producing induced finite Chichilnisky rules by repeating finite vectors, implies that $\mathcal{T}_u$ is also an upper bound on the sets of topologies able to deal with infinite Chichilnisky rules.[6] This technique considers the following embedding of $\mathbb{R}^n$ into $\mathbb{R}^\infty$:

$$\text{Rep}: \mathbb{R}^n \to \mathbb{R}^\infty : \mathbf{x} \mapsto {}_{\mathbf{rep}}\mathbf{x}.$$

and produces a sequence $F_1, F_2, \ldots, F_n, \ldots$ of induced maps defined by

$$F_n : \mathbb{R}^n \to \mathbb{R} : \mathbf{x} \mapsto F_n(\mathbf{x}) = F({}_{\mathbf{rep}}\mathbf{x}),$$

where F is an infinite Chichilnisky rule on $\mathbb{R}$. Let $\mathbb{R}^n$ be equipped with the standard topology $\mathcal{N}$ and let $\mathbb{R}^\infty$ be equipped with a topology $\mathcal{S}$. Imposition

[5] Lauwers (1993b) deals with different infinite versions of the anonymity axiom. Absence of impatience appears to be far too weak to capture the idea of equal treatment of all generations. A stronger form called bounded anonymity is introduced.
[6] Induced rules appear in Candeal et al. (1992). Efimov and Koshevoy (1992) and in Lauwers (1993a, b). The replication technique is also used in equilibrium theory (e.g. Debreu and Scarf 1963).

of the following condition upon a topology $\mathscr{S}$ requires that continuity of an infinite rule results in continuity of the induced maps.

Definition 3. A topology $\mathscr{S}$ on l_∞ is said to satisfy the repetition-condition if the following holds. Let $\mathbf{x}_1, \ldots, \mathbf{x}_k \ldots \in \mathbb{R}^n$ be a sequence of finite streams. Then the sequence $\mathbf{x}_1, \ldots, x_k \ldots$ converges to $\mathbf{x} \in \mathscr{R}^n$ wrt $\mathscr{N}$ if and only if the sequence ${}_{\text{rep}}\mathbf{x}_1, \ldots, {}_{\text{rep}}\mathbf{x}_k \ldots$ converges to ${}_{\text{rep}}\mathbf{x}$ wrt $\mathscr{S}$.

Topologies smaller than $\mathscr{T}_u$ satisfy the above condition. Note that this condition is stronger than the diagonal convergence condition mentioned in Section 3. The next theorem completes the characterization of the uniform topology $\mathscr{T}_u$ on l_∞.

Theorem 2. *Let $\mathscr{S}$ be a topology on l_∞ induced by a metric d, that allows for the technique of induced rules, i.e. $\mathscr{S}$ satisfies the repetition-condition. Then $\mathscr{S} \subseteq \mathscr{T}_u$.*

Proof. By contradiction. Therefore, assume that $\mathscr{S}$ satisfies the repetition-condition and that $\mathscr{S} - \mathscr{T}_u \neq \emptyset$. Then, there exists an open neighbourhood $O \in \mathscr{S}$ of $\mathbf{0}$ such that for all $n \in \mathbb{N}_0$, we have

$$\text{not } O_n = \{\mathbf{u} \mid d_u(\mathbf{u}, \mathbf{0}) < 1/n\} \subseteq O.$$

Let $\mathbf{u}_n$ be such that $\mathbf{u}_n \in O_n - O$. The sequence $\mathbf{u}_n$ converges to $\mathbf{0}$ with respect to $\mathscr{T}_u$. But, $\mathbf{u}_n$ does not converge to $\mathbf{0}$ with respect to $\mathscr{S}$. However, the repetition condition [even the diagonal convergence condition] demands that any sequence $\mathbf{v}_n$ satisfying $(-1/n) < \mathbf{v}_n < (1/n)$ converges to $\mathbf{0}$. This is a contradiction. Conclude that $\mathscr{S} \subseteq \mathscr{T}_u$.

The combination of Theorems 1 and 2 establishes the characterizaton of the sup topology. In view of the difficulties and the manipulability in the specification of a continuity demand, these theorems underbuilt the choice of the sup topology. It in an aggregation procedure continuity is looked upon as an appealing property, then continuity with respect to the sup topology is the only infinite version of the continuity axiom compatible with equity and able to link an infinite horizon with finite ones.

Finally, we mention that the family of $\mathscr{T}_u$-compact sets is not an interesting family. Indeed, the sets $C_1 \times \ldots \times C_n \times \ldots$ which are compact with respect to the product topology $\mathscr{T}_p$ are not compact, not even locally compact, with respect to the sup topology $\mathscr{T}_u$. Hence, in view of effectiveness, continuity with respect to the sup topology cannot be motivated.[7]

7. The Svensson topology

In his axiomatic approach to the construction of intertemporal choice functions, Diamond (1965) established the following impossibility result. A preference order on l_∞ cannot combine equity, strong monotonicity, and continuity with respect to the sup topology. Svensson (1980) introduced a weaker continuity demand and obtained a possibility result. The von Weiszäcker

[7] See Conway (1966) for $\mathscr{T}_m$-compactness.

(1965) criterion $\succsim_W$ defined by

$$\mathbf{u} \succsim_W \mathbf{v} \text{ if for some } T, t \geq T \text{ implies } \textstyle\sum_{k=1}^{t} (u_k - v_k) \geq 0,$$

can be extended to a complete preference order that is equitable, strongly monotonic, and continuous with respect to the Svensson topology $\mathcal{T}_s$. Svensson (1980, p. 1251) notes that the topology $\mathcal{T}_s$ is larger than the sup topology $\mathcal{T}_u$, but argues that $\mathcal{T}_s$ is definitely weaker than the discrete topology $\mathcal{T}_d$. In view of our results, the relevance of this topology can be questioned. We provide two arguments against the use of the Svensson topology. The first one is based upon Theorem 2, the second upon Proposition 1.

As the Svensson topology does not satisfy the diagonal convergence condition, the technique of induced rules cannot be used. Indeed, let k and n be two (large) natural numbers, and let $\mathbf{f}_n = (1/n, O, \ldots, O) \in \mathbb{R}^k$. Then, the sequence $\mathbf{f}_1, \mathbf{f}_2, \ldots, \mathbf{f}_n \ldots$ converges to $(O, \ldots, O) \in \mathbb{R}^k$, while the sequence ${}_{\text{rep}}\mathbf{f}_1, {}_{\text{rep}}\mathbf{f}_2, \ldots, {}_{\text{rep}}\mathbf{f}_n \ldots$ does not converge with respect to $\mathcal{T}_s$. A small perturbation is only a fraction of $1/k$ of the total of coordinates is with respect to the Svensson topology considered as a large perturbation.

Second, consider the following non-substitution axiom which is satisfied by the von Weiszäcker criterion.

Definition 4. A preference order $\succsim$ is said to satisfy the non-substition condition if for all $\varepsilon > 0$, we have

$$(\mathbf{u} + (\varepsilon)) \succ \mathbf{u}',$$

for all $\mathbf{u}, \mathbf{u}' \in l_\infty$ identical up to the first coordinate.

This condition expresses the idea that a large improvement in the first generation never compensates a small improvement sustained for all generations. The next lemma investigates the impact of this condition upon the representability of a preference order.

Lemma 3. *Let $\succsim$ be a strongly monotonic preference order on l_∞. If $\succsim$ satisfies the non-substitution condition, there is no real valued map F on l_∞ that represents the order $\succsim$.*

Proof. The restriction of the preference order $\succsim$ to the set of utility streams of the form $(r, (s)) = (r, s, s, \ldots, s \ldots)$ with r, s two real numbers, is a lexicographic order on $\mathbb{R}^2$:

$$(r, (s)) \succsim (r', (s')) \quad \text{iff} \quad [s > s'] \quad \text{or} \quad [s = s' \text{ and } r \geq r'].$$

And, lexicographic orders are outstanding examples of orders which are not representable by a real valued map (e.g. Tanguiane 1991, Example 2.4.2).[8]

The search for a representation of the von Weiszäcker criterion is a vain one. Imposing continuity with respect to the Svensson topology does not guarantee the existence of a representation, and cannot be motivated from this point of view.

Also, from functional analysis it follows that a linear map F on l_∞ is bounded if only if F is continuous with respect to the sup topology $\mathcal{T}_u$

---

[8] Skala (1973) showed that the lexicographic order on $\mathbb{R}^2$ can be represented by a map $F: \mathbb{R}^2 \to {}^*\mathbb{R}$, where ${}^*\mathbb{R}$ is a non-standard extension of $\mathbb{R}$.

(e.g. Peressini 1967). Continuity with respect to the Svensson topology is compatible with unboundedness, as illustrated by the von Weiszäcker criterion.

To summarize, where continuity with respect to the sup topology implies the well-behaviour of a preference order, in the sense that the order is representable (Proposition 1) and can be studied by induced rules (Theorem 2): the Svensson topology completely fails these targets as it is too large.

8. Concluding remarks

The standard topology $\mathcal{N}$ on $\mathbb{R}^n$ can be extended in different ways to a topology on l_∞. In a finite setting different motivations to impose continuity are all fulfilled by the standard topology. In this note we focussed on effectiveness, l_1-summability and a weak form of equity and showed that, in an infinite setting, each motivation is linked to another topology.

The equity demand appeared to impose the strongest condition upon the topology: only topologies larger than or equal to the uniform topology are to be considered. Dealing with intergenerational choice functions, Koopmans (1960) already makes this point. He argues that the sup topology treats all future generations alike. In addition, Koopmans (1960, note 5) meets the loss of Tychonov's theorem by means of an effectiveness postulate and shows that a set of five postulates implies an impatient ranking of utility streams. That $\mathcal{T}_u$ allows for equity, stresses the strength of Koopmans result. Diamond (1965) deals with the same framework, but considers besides $\mathcal{T}_u$ also the product topology $\mathcal{T}_p$. But, since the product topology only allows for myopic rankings, the incompatibility between $\mathcal{T}_p$-continuity, equity and monotonicity is a much weaker result contrary to what Koopmans believed (Koopmans 1972, §8).

Finally, from a mathematical point of view, the topologies $\mathcal{T}_p$, $\mathcal{T}_u$ and $\mathcal{T}_m$ are all "natural" topologies on l_∞.

First, the product topology behaves elegantly. Several useful properties are product invariant. For example, the cartesian product of compact, connected, or metrizable spaces is compact, connected, or (in case of a countable product) metrizable. Therefore, the Tychonov topology is considered the "right" topology on cartesian products, and Tychonov's theorem is one of the most powerfull single results in topology (Kelley 1995).

Second, the uniform topology $\mathcal{T}_u$ is the topology for which boundedness and continuity, when restricted to linear functionals, coincide.

Third, when dealing with linear functionals, the question whether such a map is representable by an infinite sum is a natural one. An answer of a topological nature is provided: a linear monotonic map, continuous with respect to the myopic topology $\mathcal{T}_m$, is representable by a converging infinite sum.

References

Brown DJ, Lewis LM (1981) Myopic economic agents. Econometrica 49: 359–368
Campbell DE (1985) Impossibility theorems and infinite horizon planning. Soc Choice Welfare 2: 283–293

Candeal JC, Induráin E, Uriarte JR (1992) Some issues related to the topological aggregation of preferences. Soc Choic Welfare 9: 213–227
Chichilnisky G (1980) Social choice and the topology of spaces of preferences. Adv Math 37: 165–176
Chichilnisky G (1982) Social aggregation rules and continuity. Quart J Econ 87: 337–352
Chichilnisky G. Heal G (1979) Social choice with infinite populations: Construction of a social choice rule and impossibility results. Columbia University Economics Discussion Paper
Chichilnisky G, Heal G (1983) Necessary and sufficient conditions for the resolution of the social choice paradox. J Econ Theory 31: 68–87
Conway JB (1966) The strict topology and compactness in the space of measures II. Trans Amer Math Society 72: 474–486
Debreu G, Scarf H (1963) A limit theorem on the core of an economy. Int Econ Rev 4: 235–246
Diamond PA (1965) The evaluation of infinite utility streams. Econometrica 33: 170–177
Efimov B, Koshevoy G (1992) The topology approach to social choice with infinite populations. Math Soc Sci. to appear
Epstein LG (1986a) Intergenerational preference orderings. Soc Choice Welfare 3: 151–160
Epstein LG (1986b) Intergenerational consumption rules: An axiomatization of utilitarianism an egalitarianism. J Econ Theory 38: 280–297
Fishburn PC, Rubinstein A (1982) Time preference. Int Econ Rev 23: 677–694
Harvey CM (1986) Value functions for infinite-period planning. Manag Sci 32: 1123–1139
Kannai Y (1970) Continuity properties of the core of a market. Econometrica 38: 791–815
Kelley JL (1955) General topology. Von Nostrand, New York
Koopmans TC (1960) Stationary ordinal utility and impatience. Econometrica 28: 287–309
Koopmans TC (1972) Representation of preference orders over time. In: Decision and Organization. McGuire CB, Radner R (eds.) North-Holland, Amsterdam
Lauwers L (1997) A note on weak ∞-Chichilnisky rules. Soc Choice Welfare 14: 357–358
Lauwers L (1993a) Rawlsian equity and generalized utilitarianism with an infinite population. KULeuven, mimeo
Lauwers L (1993b) Topological aggregation, the case of an infinite population. KULeuven, mimeo
Peressini A (1967) Ordered Topological Vector Spaces. Harper and Row, New York
Shinotsuka T (1994) Equity, continuity, and myopia: a generalization of Diamond's impossibility theorem. University of Rochester, working paper
Skala HJ (1974) Nonstandard utilities and the foundation of game theory. Int J Game Theory 3: 67–81
Svensson LG (1980) Equity among generations. Econometrica 48: 1251–1256
Tanguiane AS (1991) Aggregation and representation of preferences: introduction to mathematical theory of democracy. Springer, Berlin, Heidelberg
von Weizsäcker OC (1965) Existence of optimal programmes of accumulation for an infinite time horizon. Rev Econ Stud 32: 85–104

A note on weak ∞-Chichilnisky rules

Luc Lauwers

Monitoraat E.T.E.W., Katholieke Universiteit Leuven, B-3000 Leuven, Belgium

Received: 5 November 1992/Accepted: 22 April 1996

Abstract. There does not exist a weak ∞-Chichilnisky rule as defined in Candeal et al. (1992).

First, we introduce some notation. After repeating the definition of "weak ∞-Chichilnisky rules" as it appears in Candeal et al. (1992), we prove the above statement. For the interpretation and motivation of such rules, we refer to Candeal et al. (1992) and Chichilnisky and Heal (1979).

Notation. Let $\underline{X}$ be a topological space of preferences and let $\mathbf{x} = (\underline{x}_0, \underline{x}_1, \ldots, \underline{x}_t, \underline{x}_s, \ldots)$ be a vector in $\underline{X}^\infty$ [bold face letters denote elements of a Cartesian product and ∞ stands for $|\mathbb{N}|$]. For two natural numbers **x** and *s*. we denote $\mathbf{x}^{t\sim s}$ for the vector $(\underline{x}_0, \underline{x}_1, \ldots, \underline{x}_s, \ldots, \underline{x}_t, \ldots)$ obtained from **x** after exchanging $\underline{x}_t$ and $\underline{x}_s$. If there is no confusion we write $\mathbf{x}^{ts}$ for $\mathbf{x}^{t\sim s}$.

Definition. A weak ∞-Chichilnisky rule on a Hausdorff topological space $\underline{X}$ is a map

$$\Phi : \underline{X}^\infty \to \underline{X} : \mathbf{x} = (\underline{x}_0, \underline{x}_1, \ldots, \underline{x}_n, \ldots,) \to \Phi(\mathbf{x})$$

satisfying:
(i) Continuity: Φ is continuous with respect to the Tychonov or product topology on $\underline{X}^\infty$.
(ii) Finite Anonymity: For any pair $\underline{s}$, $\underline{t}$ of natural numbers and every $\mathbf{x} \in \underline{X}^\infty$, we have $\Phi(\mathbf{x}^{t\sim s}) = \Phi(\mathbf{x})$.
(iii) Unanimity: For every $\underline{x} \in \underline{X}$, we have $\Phi(\underline{x}, \underline{x}, \ldots, \underline{x}, \ldots,) = \underline{x}$.

Theorem. *There does not exist a weak ∞-Chichilnisky rule in any Hausdorff space $\underline{X}$.*
Proof. Assume that the cardinality of $\underline{X}$ is at least two and let $\underline{a}$, $\underline{b}$ be two elements of $\underline{X}$ such that $\underline{a} \neq \underline{b}$. We prove the theorem by contradiction. Therefore, let Φ be a weak ∞-Chichilnisky rule on $\underline{X}$. Consider the vectors

$\mathbf{a} = (\underline{a}, \underline{a}, \ldots, a, \ldots)$, $\mathbf{b} = (\underline{b}, \underline{b}, \ldots, b, \ldots)$ and $\mathbf{d} = (\underline{b}, \underline{a}\, \underline{a}, \ldots, a, \ldots) \in \underline{X}^{\infty}$. The sequence $\mathbf{d}, \mathbf{d}^{01}, \mathbf{d}^{02}, \ldots, \mathbf{d}^{0n}, \ldots$ converges with respect to the Tychonov topology to $\mathbf{a}$. Finite anonymity implies that for all $n \in \mathbb{N}$, we have $\Phi(\mathbf{d}^{0n}) = \Phi(\mathbf{d})$. From the continuity of Φ it follows that

$$\Phi(\mathbf{d}) = \lim_{n \to \infty} \Phi(\mathbf{d}^{0n}) = \Phi(\mathbf{a}).$$

And, unanimity of Φ implies that $\Phi(\mathbf{d}) = \Phi(\mathbf{a}) = \underline{a}$. Apply the same reasoning upon the vector $\mathbf{f} = (\underline{b}, \underline{b}, \underline{a}, \ldots, a, \ldots)$ and the sequence $\mathbf{f}, \mathbf{f}^{12}, \mathbf{f}^{13}, \ldots, \mathbf{f}^{1n}, \ldots$ which converges to $\mathbf{d}$. Conclude that $\Phi(\mathbf{f}) = \Phi(\mathbf{d})$. Let $\mathbf{D}_k = (\underline{b}, \underline{b}, \ldots, \underline{b}, \underline{a}, \ldots, \underline{a}, \ldots)$ be the vector in which the first k coordinates are equal to $\underline{b}$. Note that $\mathbf{d} = \mathbf{D}_1$ and $\mathbf{f} = \mathbf{D}_2$. By recursive use of the above reasoning it follows that $\Phi(\mathbf{D}_k) = \Phi(\mathbf{D}_{k-1})$. Hence, we have constructed a sequence $\mathbf{D}_1, \mathbf{D}_2, \ldots, \mathbf{D}_n, \ldots$ for which the images under Φ satisfy $\Phi(\mathbf{D}_n) = \Phi(\mathbf{D}_1)$ for all $\underline{n} \in \mathbb{N}$.

The sequence $\mathbf{D}_1, \mathbf{D}_2, \ldots, \mathbf{D}_n, \ldots$ converges with respect to the Tychonov topology to $\mathbf{b}$. Again, by continuity and unanimity one concludes that $\Phi(\mathbf{D}_1) = \Phi(\mathbf{b}) = \underline{b}$. Since $\Phi(\mathbf{D}_1) = \Phi(\mathbf{d}) = \underline{a}$, a contradiction follows.

Note that this impossibility theorem is stronger than Theorem 3 of Candeal et al. (1992). Examples of weak ∞-Chichilnisky rules are wrong and theorems based upon the existence of weak ∞-Chichilnisky rules lose their power. Our theorem stresses the impact of continuity with respect to the Tychonov topology in case of infinite dimensional spaces (see also Lauwers 1992). Finally, we refer to Efimov and Koshevoy (1992) where similar results are obtained.

References

Candeal JC, Induráin E, Uriarte JR (1992) Some issues related to the topological aggregation of preferences. Soc Choice Welfare 9: 213–227

Chichilnisky G, Heal GM (1979) Social Choice with Infinite Population: Construction of a Rule and Impossibility Results. Working Paper, Columbia University and the University of Essex

Efimov B, Koshevoy G (1992) The Topological Approach to Social Choice with Infinite Populations. Working Paper, Russian Academy of Sciences, Central Economics Mathematical Institute

Lauwers L (1992) Continuity in Spaces of Infinite Dimension. Unpublished Note

MIX
Papier aus verantwortungsvollen Quellen
Paper from responsible sources
FSC® C105338

If you have any concerns about our products,
you can contact us on
ProductSafety@springernature.com

In case Publisher is established outside the EU,
the EU authorized representative is:
Springer Nature Customer Service Center GmbH
Europaplatz 3, 69115 Heidelberg, Germany

Printed by Libri Plureos GmbH
in Hamburg, Germany